OPTIMIZING BANDWIDTH

THE McGRAW-HILL SERIES ON COMPUTER COMMUNICATIONS (SELECTED TITLES)

Optimizing Bandwidth

Michele Petrovsky

McGraw-Hill

New York   San Francisco   Washington, D.C.
Auckland   Bogotá   Caracas   Lisbon   London
Madrid   Mexico City   Milan   Montreal   New Delhi
San Juan   Singapore   Sydney   Tokyo   Toronto

Library of Congress Cataloging-in-Publication Data

Petrovsky, Michele.
    Optimizing bandwidth / Michele Petrovsky.
       p.   c.m. — (The McGraw-Hill series on computer communications)
    Includes index.
    ISBN 0-07-049889-X
    1. TCP/IP (Computer network protocol) 2. Intranets (Computer networks) 3. Telecommunication—Traffic. I. Title. II. Series.
TK5105.585.P48   1998
004.6'2—dc21
                                                      98-16887
                                                        CIP

McGraw-Hill

A Division of The **McGraw·Hill** Companies

1 2 3 4 5 6 7 8 9 0 DOC/DOC 9 0 3 2 1 0 9 8

ISBN 0-07-049889-X

The sponsoring editor for this book was Steven Elliot, the editing supervisor was Curt Berkowitz, and the production supervisor was Claire Stanley. It was set in Vendome ICG by Priscilla Beer of McGraw-Hill's Professional Book Group composition unit in cooperation with Spring Point Publishing Services.

Printed and bound by R. R. Donnelley & Sons Company.

 This book is printed on recycled, acid-free paper containing a minimum of 50% recycled, de-inked fiber.

This book, like all my writing, draws heavily upon the work of others. Specifically, in the case of *Optimizing Bandwidth*, I've relied on research and design from a number of levels of academia. I'd like to take this opportunity to commend those whose ideas and efforts have contributed so much to this volume. By and large, they do what they do, not for profit, but for the love of it. Their devotion to their work came across so clearly that it lent interest and excitement to mine as I developed this book's manuscript. I only hope I've been successful in conveying some of the enthusiasm for the topic that I found in their work.

CONTENTS

Contents

Contents

Contents

PREFACE

Have you ever noticed how, after struggling to absorb and understand some new data processing concept, technique, or technology, you find yourself saying "But that's just common sense/Boolean logic/modularity/whatever"? Indeed it is; there's very little new, even under this particular, seemingly singular, sun.

A book on as involved and abstract a topic as bandwidth might seem an odd place to preach the virtues of common sense. But the nature of this book's subject makes it, in my opinion, the perfect place to do just that. I think you'll find, as you go through this volume, that two seemingly commonplace conclusions form in your mind:

- the complexity of data transmission itself creates the potential for a great deal of inefficiency in the use of bandwidth

- while you can't change that complexity, you can, by carefully monitoring your network, keep bandwidth inefficiency at bay

Know what? You're absolutely right.

—MICHELE PETROVSKY

WHAT'S IN THIS BOOK

Until a few years ago, one of a PC user's most common complaints was "Why does my (print job/spreadsheet calculation/you-name-it at the application level) take so long?" Now, in this era of exponential increase in Internet use, that lament has been supplanted by something like "Why does it take so long to download this you-name-it?" Both these protestations are directly related to one of the least glamorous and least understood system and network resources—bandwidth.

This book seeks to dispel the esotericism which invests bandwidth. It attempts to do that by providing you with:

- a thorough grounding not only in bandwidth basics, but also in the constraints imposed upon its usage by such factors as communications media, network topographies and extent, and server and client applications

- tips for making the most out of the bandwidth resources you already have available

- suggestions for additional resources and tools which would be of great value to any environment

- extensive reference material on all aspects of bandwidth

What's more, *Optimizing Bandwidth* examines real-world scenarios, benchmarking and monitoring these both before and after the implementation of bandwidth tweaks, in order to offer you down-to-earth suggestions on how to configure your own bandwidth resources more efficiently.

How This Book Is Organized

Specifically, *Optimizing Bandwidth* is structured as follows:

PART 1: Bandwidth Basics

Chapter 1: TCP/IP Networks and Bandwidth
- protocols and packets
- speed and throughput

- the effect of the nature of the physical channel
- the effect of the network's physical and geographical distribution
- the effect of connectivity methods, hardware, and software
- the effect of operating systems
- the effect of applications

Chapter 2: Transmission Media and Their Effects on Bandwidth

- coaxial
- fiber data distributed interface (FDDI)
- unshielded twisted pair (UTP)

Chapter 3: Topographies and Bandwidth

Chapter 4: Connectivity Methods and Demands on Bandwidth

- dial-up
- direct connections
- satellite

Chapter 5: Connectivity Hardware and Demands on Bandwidth

- traditional modems
- ISDN modems
- multiport cards and communications servers

Chapter 6: Protocols and Demands on Bandwidth

- protocols
- user interfaces

Chapter 7: UNIX and Demands on Bandwidth

- UNIX and Linux
- Windows NT
- Windows 95

Chapter 8: Windows NT, Windows 95, and Demands on Bandwidth

- UNIX and Linux
- Windows NT
- Windows 95
- Windows 3.x

Chapter 9: Apache, Internet Information Survey, and Demands on Bandwidth

- Apache
- Internet Information Server and related applications
- Personal Web Server and related applications
- Netscape servers

Chapter 10: Client Applications and Demands on Bandwidth: A Paradigm

Who Should Read This Book

Anyone involved in the configuration, maintenance, or management of an intranet or Internet site can benefit from this volume. And even those whose interest in bandwidth extends only to answering questions like "Is there a way I can make GIFs download more quickly?" will find much useful information here (including the answer to that and other similar queries).

Conventions Used in This Book

Optimizing Bandwidth follows a few conventions, most of which pertain to how the book presents information to you.

1. Anytime we use a technical term for the first time, it will be presented in full, and in italic, and followed immediately by an acronym for the term.

2. Each chapter in *Optimizing Bandwidth* begins with a short description of the chapter's topic.

3. Every chapter ends with two sections:

   - In a Nutshell, which contains a brief review of the most important concepts in the chapter
   - Looking Ahead, which gives a thumbnail account of what the following chapter will present

ACKNOWLEDGMENTS

Many, many people helped bring this book to life; it would be impossible to thank them all. But I'd like to mention three whose help was especially important:

- Peter Druschel of Rice University's Computer Science department, for his quick and informative responses to my questions-by-email

- Steve Elliot, for considering the original proposal worth reviewing, and for the many enjoyable title- and other design-brainstorming sessions

- Tommy, for acquiescing as always to my many requests to "read this and tell me what you think"

—MICHELE PETROVSKY

OPTIMIZING BANDWIDTH

Bandwidth
Basics

TCP/IP Networks and Bandwidth

In this first chapter, we review:

- the makeup of TCP/IP and TCP/IP networks
- the effects of this ubiquitous protocol and hardware suite upon bandwidth consumption

A Client/Server Model

The TCP/IP protocol suite, the keystone of the Internet, is built upon a client/server model. Under such a paradigm, one computer or subsystem, the *server,* provides services; another, the *client,* must specifically request those services.

A client always initiates the conversation by issuing such a request. A server, on the other hand, does nothing more than sit patiently waiting for client petitions.

NOTE: *This server passivity is known in the lingo of data communications as being in a* wait state.

Exchanges between client and server rely on messages, which in the world of TCP/IP have a very specific nature and structure. (We examine TCP/IP messages in more detail later in this chapter.) When a server receives a message that contains a request for services from a client, the server first finds out if those services are available. Then it determines if the client has been allowed access to the services. Finally, the server figures out how it will deliver the services to the client.

TCP/IP in More Detail

TCP/IP, once the default protocol suite only of UNIX systems, has become the norm on the Internet at large. As one might expect, given this omnipresence, TCP/IP happily ignores:

- a computer's manufacturer, and therefore can be the basis for heterogeneous networks
- the physical layout of the network
- the operating systems involved in a network

Rather, TCP/IP forces both client and server to rely on a single standard.

The TCP/IP suite includes not only protocols, like the *transmission control protocol* (TCP) and *Internet protocol* (IP), for which it is named and which function at relatively low levels of the *open systems interconnection* or OSI model, but also application-level protocols that accomplish, among other things:

- email
- emulation
- file transfer
- remote login

Signals

Now that we've taken a look at the outermost level of TCP/IP, let's examine it from the inside out. We begin with a review of the nature of electromagnetic signals.

Such signals can be classified as either *continuous* or *discrete*, that is, as *analog* or *digital*. In an analog signal, intensity varies in a smooth, flowing way over the life of the signal. With digital signals, on the other hand, intensity maintains a constant level for each time period during which the signal exists. Further, that intensity can, in most cases, occupy only one of two possible levels.

Both analog and digital signals can be added together, or more correctly placed side by side, to produce a larger, composite signal. When dealing with such composite signals, two characteristics come to the fore:

- *Spectrum:* the *range of frequencies* the composite contains
- *Bandwidth:* the *range or width of the signals' spectrum*

Digital Signals. Since computers rely entirely upon digital signals to transmit data, we confine the rest of our discussion to this group.

Digital signaling for data transmission uses electromagnetic pulses that represent the two binary digits 0 and 1. In this scheme, a constant positive voltage indicates 0, whereas a constant negative voltage signifies 1. So, information of any sort must first be converted to binary form (if that hasn't already happened) before it can be transmitted. This conversion can therefore be considered a corollary of bandwidth usage, since the speed and way in which it takes place can affect that usage.

Digital Signals and Bandwidth. In order to understand the most significant effect of digital signals on bandwidth consumption, we first take a moment to illustrate those signals, in Fig. 1-1.

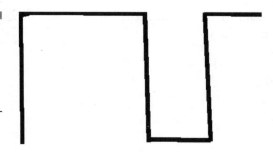

Figure 1-1
The ideal for digital signals like this one has been described as "nice sharp corners."

An idealized digital signal like that in Fig. 1-1 would require *infinite* bandwidth, in order to ensure those nicely squared corners, and what they represent—the ability to completely accurately convey information. Although few environments can provide such bandwidth, most can hope to accomplish a more realistic means of ensuring transmission capacity and accuracy. That is, most environments can aspire to providing a *range of frequencies,* or *bandwidth,* which is large enough to:

■ accurately represent a stream of digital pulses

■ provide adequate information-carrying capacity

Unfortunately, the converse is too frequently the case in the real world. With bandwidth often significantly limited, there is not only scarcity but appreciable potential for signal distortion, and therefore for error, at the receiving end.

Effects Upon the Accuracy of Signals. Beyond sheer bandwidth available, there are a number of factors which can affect the condition in which a digital signal arrives at its destination. We investigate the most important of these now.

ATTENUATION. Any electromagnetic signals will weaken the longer they must travel. In order to counter such *weakening of signals over distance,* or *attenuation,* networks use devices called *repeaters,* which, as Fig. 1-2 depicts:

■ are placed at predefined intervals along signal paths

■ reconstitute signals before allowing them to proceed further along their paths

Figure 1-2
Repeaters are most effective when positioned to act as hubs too.

NOTE: *Attenuation takes place more readily at higher frequencies, and thus has a greater distorting effect on messages which travel that part of a bandwidth spectrum.*

DELAY DISTORTION. Another frequency-related source of distortion for digital signals is a phenomenon that occurs in some categories of transmission cables. At different frequencies, the velocity of propagation of such signals through a cable can also differ. This *variability of velocity of a digital signal,* in turn, can result in a type of distortion called *delay distortion.* Delay distortion can introduce significant errors into digital data, since it can contort voltage levels, and therefore the 0s and 1s those levels represent.

NOISE. Noise related to the nature of a transmission medium rather than to that of a signal, must nonetheless be taken into account in any discussion of bandwidth. In order to compensate for noise and the errors it can introduce into digital data transmissions, extra bandwidth may be exhausted.

Noise has been defined as unwanted electromagnetic energy that is inserted somewhere between transmission and reception. Four broad categories of network noise exist. We've summarized these in Table 1-1.

TABLE 1-1

Network Noise

Type of Noise	Source
Crosstalk	Spontaneous coupling of signal paths, such as that which takes place when signals transmitted by noncable methods overlap.
Impulse noise	Irregular pulses, or *spikes*, of short duration and relatively high amplitude. Impulse noise most frequently results from flaws in the transmission system; it is the most significant cause of error in data transmissions.
Intermodulation noise	Signals of different frequencies sharing the same transmission medium. Intermodulation noise results in signals occupying a frequency that is the difference, multiple, or sum of the original frequencies.
Thermal noise	Heat-related agitation of electrons in a conductor. Thermal noise occurs in all electronic devices and transmission media, and can never be completely eliminated.

Transmission Rates and Bandwidth Capacity. Of course, any transmission medium, no matter how well designed or maintained, can offer only finite bandwidth. The precise amount available is a function of:

- limitations of the transmitting system (the server)
- measures intended to prevent distortion
- the physical properties of the medium

The most important relationship in data transmission is that between bandwidth available and transmission rates. This relationship is approximately 1 to 1; so, doubling bandwidth nearly doubles the rate at which data moves across a network. However, noise must be taken into account in determining the net rate at which reliable data arrives at a client. As data transmission rates increase, an effect similar to those related to frequency discussed earlier in this chapter takes place. At higher transmission rates, the portion of the signal stream occupied by individual bits becomes smaller, making more bits subject to distortion by noise. It's for this reason that signal-to-noise ratios must be considered when estimating net transmission rates. One of the fundamental equations of information theory, Shannon's equation for channel capacity, allows us to do just that. Figure 1-3 presents and explains Shannon's equation.

Figure 1-3
Shannon's equation assumes that noise in signal-to-noise ratios is thermal.

$$C = W (LOG\_2) (1 + S/N)$$

where C = channel capacity, W = raw bandwidth, and
S/N = signal to noise ratio

NOTE: *In practice, only much lower rates than those indicated by Shannon's equation are achieved.*

If we mull over Shannon's equation, we see that net data transmission rates, that is, transmission efficiency, can be increased by increasing either:

- signal strength
- bandwidth

But, as noted earlier in this section, increasing bandwidth also introduces greater opportunities for signal distortion. Are we, then, left with the prospect of slowly but steadily decreasing returns in transmission efficiency? No; the better solution is to improve the use of bandwidth resources currently available.

TCP/IP Component Areas

Any system or network that adheres to the client/server model contains a number of subsystems. These fall into the following categories:

- client components
- server components
- middleware
- connectivity hardware

Client Operating System. A client computer must rely on its native operating system for its means of access to network services. The client OS handles both internal processing related to data communications, and more important, the interface to middleware, that is, connectivity software and protocols. Thus, the client operating system has a significant effect on the overall efficiency of data communications.

Server Operating System. Like client operating systems, server OSes handle local operations related to data communications. Unlike their client counterparts, though, server operating systems must handle not only interprocess communications but also the applications that make up a large part of the resources distributed by any network.

Connectivity Software and Protocols. Middleware is the TCP/IP component area that is by far the most critical to data communications. This category contains all the software needed to establish and manage the client/server conversation. Middleware includes:

- network operating systems or NOSes
- service-specific software such as device drivers
- transmission software and protocols

Table 1-2 identifies the characteristics of each of these varieties of middleware that are most important to discussions of bandwidth.

THE MAXIMUM TRANSFER UNIT. Any message traveling a TCP/IP network consists of *transfer units* or *frames.* In such a network, the *largest frame size* usually available, or *maximum transfer unit* (MTU), is 1500 bytes.

NOTE: *Backbone or essential network segments have been cabled more and more in recent years with fiber digital data interface or FDDI channels. FDDI's MTU is about 4350 bytes. FDDI can do more than sling data out faster than metal cables. It's also* hardly *subject to data distortion due to noise.*

	Middleware Category	Bandwidth Affected Through
TABLE 1-2		
Middleware and Bandwidth	NOSes	Coordinating the execution of network applications, and the allocation of resources to those applications.
	Service-specific software such as email	Presenting requests for additional connections
	Transmission software	Providing protocols that direct the flow of network traffic.

All IP-based messages must adhere to these MTUs. Any message or packet larger than the applicable MTU will be divided and sent as separate, independent packets that must be reassembled upon reaching their destination. In heterogeneous networks combinations of MTUs can be present, forcing IP packets to be reduced to the lowest possible denominator. This yardstick for network messages is called a *path MTU*.

DIAL-UP PROTOCOLS.   At home and even to some extent at work, most of us use modems to reach the Internet. These modems connect to carriers through an RS-232 serial port, notorious for a tortoise-like pace. In general, serial communications drastically reduce data transmission throughput. Two protocols attempt to address this problem: the *serial line Internet protocol* or SLIP, and the *point-to-point protocol* or PPP. Let's examine these in more detail.

SLIP.   SLIP simplifies data transmissions by assuming that the stations at either end of a conversation know each other's IP address. SLIP thereby eliminates one of the fields of a typical TCP/IP message header, the destination address. In addition, SLIP forgoes another such field, the *cyclic redundancy check* or CRC, and assumes instead that error detection will be handled by higher OSI layers. For these and other reasons, SLIP cannot be combined with any other TCP/IP protocol during a session it has established.

CSLIP.   Despite their frugality in some areas, traditional SLIP transmissions still include a 20-byte IP header, and either a 20-byte TCP header or 8-byte *user datagram protocol* (UDP) header. So, SLIP can experience significant transmission delays, which can in turn be exacerbated if a session is highly interactive. As an alternative to these SLIP flaws, the *compressed serial line Internet protocol* or CSLIP was developed. CSLIP substitutes a scrawny 5 bits for SLIP's 40-byte TCP/IP headers, and an even skinnier 3 bits for its 28-byte UDP/IP ones.

PPP.   Like SLIP, PPP can run on simple serial lines. Like CSLIP, PPP compresses TCP and IP header information. But unlike SLIP, PPP does more; it offers:

- a 2-byte CRC field at the end of each frame, allowing error detection
- address advertising and resolution, thereby expediting routing
- a multiprotocol capability, which allows PPP to use network-level

protocols other than IP, making PPP more efficient in heterogeneous environments

Connectivity Hardware.   Of the four most important types of connectivity devices:

- repeaters
- bridges
- routers
- gateways

bridges and routers have special significance to bandwidth consumption. Next, we investigate these devices and their effect on bandwidth.

ROUTING AND BRIDGING.   Routing controls the path taken by TCP/IP
messages as they travel from source to destination on a network. At
least one and often several intermediate nodes may be traversed along
that path. Bridging at first glance seems indistinguishable from routing. However, because they function at different levels of the OSI
model, bridging and routing—although they each contribute to moving data from source to destination—use different information in
doing so.

Bridging occurs at the OSI data link layer. This level of the model:

- controls data flow
- detects transmission errors
- manages access to the physical transmission medium

but does not concern itself with such internetwork or intersegment
questions as:

- original source or ultimate destination
- resolving, that is, translating between differences in protocols
- route optimization

Bridges, relatively uncomplicated pieces of hardware:

- analyze incoming frames
- make forwarding decisions based on information contained in
  those frames
- forward the frames in the direction of their destination

Sometimes, all source-to-destination path information accompanies every frame transmitted. But in many cases, frames are forwarded one hop at a time toward their intended recipient. Because bridges operate at the link layer, rather low on the OSI hierarchy, they can and do ignore addressing, routing, and other information that pertains to upper layers of the model. So, bridges also can and do forward traffic that relies on more than one type of upper-level protocol. But at the same time, bridges retain the ability to distinguish what those upper-level protocols are, and to filter the data they forward based upon that distinguishing. For example, some bridges can be configured to reject all multicast packets.

In effect, by dividing networks into semiautonomous entities, bridges can help to improve data transmission rates in several ways. Because not all traffic arriving at a bridge is forwarded, the bridge can reduce the overall traffic load on the segments it connects. Once again because of their filtering abilities, bridges can act as firewalls of sorts, for instance, rejecting all packets originating at a given source. Finally, bridges can extend the effective length of a LAN without increasing, or at least with only minimal increase in, the demands placed on bandwidth.

One category of bridges, called *remote bridges,* which connect multiple network segments occupying different physical or geographical areas, must also deal with a basic difference between *local area networks* (LANs) and *wide area networks* (WANs). Because of their more limited physical expanse, the speed of transmissions on LANs greatly exceeds that usually found on WANs, sometimes by several orders of magnitude. Although they cannot specifically improve WAN speeds, remote bridges can compensate for speed discrepancies, with buffering. For example, a LAN-based server that can transmit at rates of 10 megabits per second (Mbps) can communicate transparently with a WAN, which it must reach by means of a 64-kilobits-per-second (Kbps) serial link, if the bridge between the two networks temporarily stores data it receives, sending it on to the serial link only intermittently and at rates which that link can handle.

Routers, unlike bridges, cannot ignore addressing and routing information—far from it. Such data is a router's bread and butter. To see why this is so, take a look at the following thumbnail sketch of how routers function:

- Packet A arrives at Router Z.
- Router Z reads Packet A.
- Router Z strips off Packet A's protocol-related header.

- Router Z replaces the discarded message header with protocol-related information appropriate to the message's destination network, segment, or next stopover.

- Router Z adds path-related information, possibly including a calculated-on-the-spot optimum route, to the message header.

- Router Z sends Packet A on its way.

- Packet A's next stop is Router Y.

- ...and so on....

It should be clear from this scenario that, in addition to everything else they must do, routers have to communicate with one another, in order to be able to determine and offer optimum paths. This communication—that is, the distribution to all routers on a network of information regarding traffic conditions on that network—is important and specialized enough to require its own protocol category, *routing protocols*. Widely used TCP/IP routing protocols include:

- the *border gateway protocol* (BGP), which handles interdomain routing. (Domain in this context refers to Internet domains, not to those in the Windows NT or 95 worlds.)

- the *interior gateway routing protocol* (IGRP), which can route messages between subnetworks within an overall network; to networks outside that to which the router using IGRP belongs; and to networks served by routers that are members of the group into which the router running IGRP has been configured.

- the *open shortest path first* (OSPF) protocol, which was designed specifically to provide intersegment and internetwork routing for IP networks.

Another thing's probably clear by now. Routers had better find efficient paths for the messages they transfer—it's the least they can do, considering the amount of bandwidth overhead their protocols and router-to-router conversations introduce into a network.

The OSI Model and TCP/IP

In order to understand TCP/IP networks fully, the protocol suite must be examined from still another point of view: its relationships to the OSI model. This section does just that.

TABLE 1-3

Getting to Know
OSI

OSI Layer	OSI Model Number	Handles
Physical	1	The physical connection between a network node, be it PC, mainframe, printer, multiplexer, or the like, and the network transmission medium
Data Link	2	The movement of data from the originating hardware to the transmission medium (most frequently, a cable) by means of frames; the formatting of data to be transmitted into units, which contain not only that data but also header, trailer, and address information. The data link layer is further divided into two sublayers: *media access control* (MAC) and *logical link control* (LLC). In essence, the MAC sublayer handles source and destination network designations, whereas the LLC sublayer takes on tasks like error control, flow control, and framing for messages.
Network	3	The movement of packets within and between networks.
Transport	4	End-to-end connectivity between networks. The transport layer controls data flow for distributed or networked applications. TCP/IP can use either of two transport layer protocols: TCP and UDP. The former provides end-to-end data reliability; the latter cannot ensure that data will arrive at its destination in the form in which it left its source.
Session	5	Establishing and managing client/server conversations.
Presentation	6	Interpreting text and image formatting.
Application	7	Details of an application. Correlates its activities with those of the session and presentation layers. Includes the software and protocols for such end-user applications as the *file transport protocol* (FTP) and the *simple network management protocol* (SNMP).

OSI Layers. The OSI model consists of seven layers, numbered in ascending order from the closest to the furthest-removed from connectivity hardware. Table 1-3 sums up the OSI model's layers.

Software that follows the OSI model creates a client application request. That request takes a round trip down through and back up the model's seven layers in being fulfilled by a server. At each of the request's 14 stops, it receives its analog to a stamp on a passport—encoding that allows processes functioning at each OSI layer on the client to communicate with processes working at that same layer on the server. These several stages of encoding are what we refer to as *protocols*. Each, and the encoding process as a whole, has its effect on bandwidth consumption.

THE NETWORK LAYER IN MORE DETAIL. The role of the *Internet protocol* (IP), the TCP/IP suite's OSI network layer protocol, can be said to extend past that layer, since all TCP/IP transport layer protocols move through a network as IP datagrams. Such units are each treated as an independent entity, allowing each to be routed through a different path. However, there's no guarantee that an IP datagram will ever reach its destination. IP tries its best, but problems like routers running out of buffer space can cause it to throw away a datagram. In other words, the way in which IP operates can create bandwidth overhead.

IP Addressing. In trying its best, IP relies on IP addresses, which, whether they indicate source or destination, consist of 32 bits, presented in four sections separated by the period (.) character. Table 1-4 describes the significance of these *IP address segments*, or *octets*.

TABLE 1-4

The Anatomy of an
IP Address

For an IP Address Such as 207.103.113.182, the Octet(s)...	Mean...
207.103.113	Network size/type, address; in this case, a Class C network, or one made up of about 250 machines
182	Host address

For an IP address such as 190.103.113.182, the octet(s)...	Mean...
190.103	Network size/type, address; in this case, a Class B network, or one made up of as many as 65,000 machines
113.182	Host address

For an IP address such as 126.103.113.182, the octet(s)...	Mean...
127.	Network size/type, address; in this case, a Class A network (very rare), or one made up of as many as 16,500,000 machines
103.113.182	Host address

TABLE 1-5

The Anatomy of an
IP Message Header

Field	Number of Bits	Indicates
Version	4	The IP version being used
Header length	4	Most commonly, a value of 20 bytes; used by routers to strip off and repackage the IP header
Type of service	4	How upper-layer protocols will handle the message; analogous to a level of importance
Total frame length	2	Length of the message, including both header and actual data, in bytes; used to determine data length (by subtracting header length), to derive MTU parameters, and by routers to decide if a message must be divided
Identification	16	A serial number for each message; used for sequencing
Time-to-live	8	The number of hops or metrics the message may take; usually set to 32 or 64, decremented by one for each router a message passes through, and, when zero, causes the message to be discarded, thereby preventing infinite looping of undeliverable messages, and the bandwidth black holes such looping can produce
Source and destination addresses	32	Where the message originated, and where it's going
Option	1	Handling options; includes possible values 1 (security restrictions), 2 (router recording), 3 (router timestamp), 4 (specification of routers that the message may employ), and 5 (a less limiting means of router specification)

The IP Message Header.   Having dissected IP addressing in Table 1-4, let's turn to Table 1-5 to do the same for the most important pieces of IP message descriptive information, that is, for the most significant components of an *IP message header.*

IP Routing.   Read this description of IP routing with the topics

■ bandwidth consumption

■ the use of network and server resources as a whole

firmly in place in your mind.

IP routing specifies that messages travel through and between networks one hop at a time. As it begins its trip, the message has no idea of its entire route. Rather, at each stop, the next interim destination is calculated, by comparing the message's destination address with an entry in a routing table. *Routing algorithms* use a *metric,* that is, a path length or hop count, to determine the most efficient path to a destination. These same algorithms set up and maintain routing tables, the repositories for network and internetwork route information. Routing algorithms attempt to provide a router with the ability to:

- select the optimal path for a message
- function efficiently, with a minimum of memory and CPU overhead
- perform correctly even under circumstances like high loads

Sophisticated routing algorithms can base route selection on multiple metrics, combining them in a single hybrid. Some of the metrics used include:

- path length
- path load
- path reliability

THE TRANSPORT LAYER IN MORE DETAIL. The transport layer of the TCP/IP suite provides end-to-end connectivity, and services the application layer. As noted earlier, TCP/IP actually contains two transport-layer protocols: TCP and UDP. Both these protocols rely on ports.

In the TCP/IP world, ports are not physical but rather virtual connections, representing the points at which upper-layer source and destination processes receive transport-layer services. Although they may be assigned as needed, many *well-known* (the actual term used in the literature, intended to indicate that these are *not required*) TCP/IP ports exist. Applications therefore frequently use:

- the numbers that represent these well-known ports
- a system-maintained table of port numbers and corresponding services

to converse with and receive transport and network layer services. Frequently used well-known TCP ports include those outlined in Table 1-6.

Communications in and out of TCP/IP ports carry a *unique serial*

TABLE 1-6

Well-Known TCP/IP Ports

Well-known Port Numbers	Most Common Use
17	Quote of the day
20	FTP data
21	FTP control
23	Telnet
25	Simple mail transfer protocol (SMTP)
41	Graphics
53	Domain Name service
137	NetBIOS Name service
156	SQL service

number called a *socket*; that's right—sockets, like TCP ports, are virtual rather than physical. Socket numbers combine the relevant IP address and port number.

TCP. TCP provides full-duplex service to upper-layer protocols. It moves data in a continuous, unstructured byte stream within which individual bytes are identified by *sequence numbers*. The TCP sequence number field usually specifies a number assigned to the first byte of data in the current message. But in some circumstances, the sequence number field can also be used to define a sequence number applicable to the entire transmission. TCP headers also contain an *acknowledgment number field,* representing the next byte of data that should be received; the *data offset field,* specifying the number of 32-bit words in the TCP header; a *flags field,* which contains control information; a *window field,* indicating the size in bytes of the buffer space available for incoming data; a *checksum field,* which tells TCP whether the header was damaged in transit; an *urgent pointer field,* which points to the first, if any, urgent data byte in the packet; an *options field,* which specifies various TCP options; and at long last, the *data field.* Suffice it to say that the structure of a TCP packet introduces still more bandwidth overhead.

UDP. UDP is a much simpler protocol than TCP, having in its header only four fields: source port, destination port, length, and checksum. The *source port* and *destination port fields* serve the same functions as they do in the TCP header. The *length field* specifies the aggregate length of

UDP header and data. The *checksum field* allows packet integrity checking, and is optional. Clearly, UDP carries much less bandwidth baggage than TCP; that's what makes UDP a better choice for interactive sessions.

THE APPLICATION LAYER IN MORE DETAIL.   The Internet protocol suite includes many upper-layer protocols representing a wide variety of applications, including network management, file transfer, distributed file services, terminal emulation, and electronic mail. In this section, we consider two of the most important.

File Transfer Protocol (FTP).   *FTP* is a utility used to distribute files in and across networks without establishing a remote logon session. Unlike another application-level member of the TCP/IP suite, the network file system or NFS, FTP does not accomplish file access, but rather only file transfer.

As Table 1-6 pointed out, FTP uses two TCP port connections: 20 for data transfer and 21 to manage that transfer. Therefore FTP can simultaneously send commands and data between a client and server. FTP conversations begin in classic client/server fashion, with the server in a wait state until the client opens port 21 to set up the control connection. This connection stays open during the entire client/server communication, while the FTP data connection on port 20 is established or reestablished every time a file is transferred. The overall FTP connection is established when the client indicates the desired host's IP address or host name.

FTP commands, and the replies by the server to those commands, are sent across the control connection. FTP's internal commands are four-character ASCII strings; replies are three-character ASCII strings. Internal FTP commands provide for connection, password verification, and options affecting the actual file transfer.

Simple Network Management Protocol (SNMP).   *SNMP*, TCP/IP's most widely used management protocol, is indeed relatively simple, but nonetheless provides tools sufficient to handle the varied and often thorny problems that can crop up in heterogeneous networks. SNMP is an application-layer protocol designed to facilitate the exchange of management information between network devices. By monitoring SNMP data, such as packets per second and network error rates, administrators can eyeball and tweak many aspects of network performance, including bandwidth consumption.

SNMP relies on what it terms *agents*, that is, software modules that run in managed devices. SNMP agents gather information about those devices and distribute this information via SNMP.

Any node on a network, including hosts, communication and other servers, printers, connectivity devices, and more, can be an SNMP-managed device. In dealing with managed devices, some of whose ability to run agents may be limited (e.g., printers with limited onboard memory), those agents attempt to minimize their impact on the performance of the managed device.

For this reason, SNMP also relies on a special type of network node, most frequently a high-powered, engineering-workstation-caliber computer, which it designates a *network management system* (NMS). NMSs run the agents that actually present management information.

Communication between managed devices and NMSs might include any or all of the following:

- the number of certain kinds of error messages transmitted
- the number of bytes and packets in and out of a managed device
- the maximum output queue length in a given period of time for routers and other internetworking devices
- the identity of network interfaces that are either going down or coming up

Managed devices will respond to four different types of commands from an NMS:

- *read,* which allows the NMS to read various status and environment variables maintained by the managed device
- *write,* which allows the NMS to change the value of such variables
- *transversal operations,* through which an NMS determines the specific variables a particular managed device supports, and gathers information from those variables for use in such system tables as IP routing tables
- *traps,* through which an NMS can cause a managed device to report specified events to the NMS

Every SNMP implementation includes a *management information base* (MIB), essentially a database of all managed devices. Every MIB is made up of fields that adhere to one of the data types:

- *Network address:* A 32-bit IP address.

- *Counter:* A nonnegative integer that increases automatically until it reaches a previously defined maximum value, at which point it returns to zero. One example of a counter might be the total number of login attempts per day from a particular remote site.

- *Gauge:* A nonnegative integer that can increase or decrease, but which requires no definition of maximum or minimum value. Example: the length of an output packet queue.

- *Tick:* Hundredths of a second since some event, as in, for instance, ticks since an FTP session was initiated.

SNMP itself is a simple request/response protocol. Nodes can send multiple requests without having received a response to any. Requests to SNMP may be for any of the following:

- *Get,* which retrieves an MIB value from an agent

- *Get Next,* a traversal operation that reads sequentially through an MIB from within an agent

- *Get Response,* which retrieves an agent's response

- *Set,* which establishes a value in the MIB

- *Trap,* used by the agent to asynchronously inform the NMS of some event

Although there's more we could tell you about SNMP and its operations, we think you have the picture by now. Like FTP and its other application-level TCP/IP peers, SNMP can add to the demands for bandwidth on a network.

IN A NUTSHELL

1. The nature of TCP/IP networks imposes constraints upon, and adds overhead to, bandwidth consumption.

2. Among the characteristics of TCP/IP that add bandwidth overhead are:

   - its reliance on a client/server model, and the practice in that model of a server's remaining in a wait state

- the inevitability of noise, and particularly of thermal noise, in any transmission
- the fact that signal-to-noise ratios affect net bandwidth available
- the fact that some protocols, like SLIP, not only must work with very slow serial lines but also offer no compression
- the need for a variety of identifying information in a TCP/IP message header
- the need for a variety of routing protocols when routers are used to connect portions of heterogeneous networks
- the need to establish and maintain virtual points of connection known as TCP/IP ports, so that protocols inhabiting various levels of the OSI model may communicate with one another
- the need of application-level protocols like FTP to encode their instructions
- the need of management tools like SNMP to offer a variety of parameters

LOOKING AHEAD

In Chap. 2, we investigate in more detail the demands placed by physical data transmission media on bandwidth usage.

CHAPTER 2

Transmission Media and Their Effects on Bandwidth

In this chapter, we investigate the constraints imposed by the most common data communications media upon bandwidth. We pay particular attention to those types of Ethernet cable:

- copper coaxial
- unshielded twisted pair

which you're most likely to use on the bulk of your network, and which are most subject to such bandwidth-busters as attenuation.

We begin by fleshing out the physical characteristics of these two typical transmission types. Then we present performance parameters for each.

10 Base 2 and 10 Base T

The *Institute of Electrical and Electronics Engineers,* or IEEE (the acronym pronounced "I Triple E") develops and disseminates standards for nearly everything under the data communications sun, including transmission media. Among the most significant of such standards are those for Ethernet cable, the most commonly used varieties of which are:

- 10 Base 5 or, colloquially, thick Ethernet
- 10 Base 2 or thin Ethernet
- 10 Base T or unshielded twisted pair Ethernet

The phrasing *X Base Y* is a notation, devised by the IEEE, whose parameters respectively indicate *three characteristics of any network cabling:*

- data transfer rate
- transmission technique; in the case of any of the cables we discuss, *baseband,* meaning that *only one signal* can use the transmission medium at any one time
- segment length of a cable of the indicated type; not applied consistently, as evidenced by the T in 10 Base T

Let's summarize common Ethernet standards in Table 2-1.

NOTE: *The IEEE did not develop the original standard for Ethernet; that set of specifications, published in 1980 by representatives of Digital Equipment Corporation, Intel Corporation, and Xerox, is sometimes referred to as the DIX standard. But beginning in 1985 with the first of several revisions to the Ethernet standard, the IEEE took over the role of Ethernet monitor and mandator. It was the document they published in 1985, whose official title is "IEEE 802.3 Carrier Sense Multiple Access with Collision Detection (CSMA/CD) Access Method and Physical Layer Specifications," which defined many of the transmission media characteristics we've come to take for granted. What is CSMA/CD? All versions of Ethernet use this media access control (MAC) protocol to manage which devices can transmit data to the network, and when they can do so. Carrier sense multiple access with collision detection could be rephrased as "Don't speak while someone else is speaking, but if it happens, everyone stop and try again later." Needless to say, such a paradigm can have negative effects on bandwidth usage.*

TABLE 2-1

IEEE Standards for Widely Used Ethernet Varieties

Cable Type	Consists of	Can Transmit at	Segment Length	Found in
10 Base 5 (thick Ethernet)	A 50-ohm copper coaxial cable. (An *ohm* is a measurement of the resistance of a medium to electronic signal)	10 megabits per second (Mbps)	500 meters	Networks that adhere to a bus topology, particularly as a backbone, as shown in Fig. 2-1
10 Base 2 (thin Ethernet)	50-ohm copper coaxial cable	10 Mbps	185 meters (go figure—it's not 200). The thinness of 10 Base 2 gives it less shielding from noise and fewer copper wires that can act as signal paths, and thereby also gives it a shorter segment length than 10 Base 5.	Cable runs between network nodes, as illustrated in Fig. 2-2
10 Base T (unshielded twisted pair, a tweak of standard telephone wire, added to the IEEE 802.3 specifications in 1989)	Wire pairs twisted about one another through the length of the cable	Anywhere from 1 to 100 Mbps, depending on the particular variety of UTP. Those are: ■ Category 2; used primarily for telephone wiring and transmitting at up to 1 Mbps ■ Category 3; also sometimes called voice-grade UTP, despite its being used primarily in Ethernet segments, which can transmit at up to 16 Mbps	100 meters for Category 3, the most significant variety	

(*Continued*)

TABLE 2-1

IEEE Standards for Widely Used Ethernet Varieties (*Continued*)

Cable Type	Consists of	Can Transmit at	Segment Length	Found in
		■ Category 4; used in token-ring topologies, and transmitting at up to 20 Mbps ■ Category 5; which further tweaks traditional UTP cable, rather than substituting the cleaner but much more expensive FDDI variety, to achieve transmissions up to 100 Mbps		A variety of settings
10 Base F	Spun-glass "wires" or fibers, which transmit pulses of light rather than blips of electrical current, and thereby incur little if any attenuation or crosstalk, although still encountering some thermal noise	10 Mbps	From 500 to 2000 meters, depending upon whether or not repeaters enter the picture	Backbones

Now that we've got a grasp of just what Ethernet cable is, let's turn to dissecting its two most widely used varieties.

10 Base 2

10 Base 2, a less-expensive version of 10 Base 5, uses a lighter and thinner coaxial cable and can do without external transceivers like those needed by Thicknet.

NOTE: *Any Ethernet cabling must work in tandem with a number of hardware components. The most significant of these in the context of thin Ethernet are:*

- *a medium dependent interface or MDI, the actual cable-to-cable or cable-to-device connectors, such as Bayonet-Neill-Cancelman or BNC Connector, the T-shaped connector so often seen in Ethernet environments*
- *a transceiver, also called a medium access unit or MAU, a tiny "black box" that actually transfers signal from computer to transmission medium or vice versa*

10 Base 2 most commonly connects segments of an overall bus topology, in which every network node attaches to the network with a BNC connector. Every segment, and the backbone, of a bus-based network must end in a 50-ohm terminator, in effect a plug that closes the circuit, that is, creates a complete, uninterrupted path for the signal.

Every node in a 10 Base 2 network must be a minimum of 0.5 meters, about 1.5 feet, from its nearest neighbor. As noted in Table 2-1, no segment cabled with 10 Base 2 can be more than 185 meters, or about 60 feet, long. Figure 2-1 depicts a simple 10 Base 2 network that adheres to these requirements.

Figure 2-1
Thinnet's cable run length limitations constrain the topologies that may be based upon it.

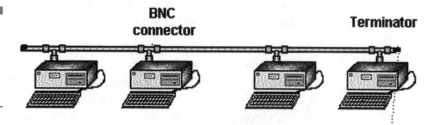

185 meters maximum cable length

Figure 2-1 illustrates another characteristic of 10 Base 2. There's no getting around the need, in such environments, for individually connecting every node to the medium. For instance, you couldn't simply string cable between a node and some other BNC connector on the network. Doing so would almost certainly disrupt transmissions, if not bring them to a halt.

Like any technology, 10 Base 2 Ethernet has inherent advantages and disadvantages. Let's examine these in a bit more detail, since they have a decided effect upon network design and management.

10 Base 2: Advantages.   On 10 Base 2's plus side—it:

- contributes to simplicity in network architectures, since 10 Base 2 networks are simply daisy-chained together with coax cable and T adapters, and generally need no hubs, transceivers, or other similar devices

- is probably the most inexpensive network-cabling technology available

- offers appreciable immunity from electrical noise caused by outside sources

10 Base 2: Disadvantages.   We've had the good news. Now let's look at 10 Base 2's down side. This cable type can be:

- difficult to reconfigure, requiring at least some down time, since, for example, new sections must be spliced in, mandating in turn cutting the existing cable at some point

- highly fault-intolerant; in other words, if any device or cable section on the network fails, the entire network may very well go down

- very hard to troubleshoot; there's just no easy way to isolate faulty cable sections or nodes

As you've probably already deduced, 10 Base 2 works most effectively in small networks that don't change frequently. One such application, illustrated in Fig. 2-2, might include multiple 10 Base T segments, and use hubs. Such a configuration allows the designer to take advantage of 10 Base 2's strengths, while minimizing one of its drawbacks: the relatively short segment length that might otherwise require the extensive use of repeaters.

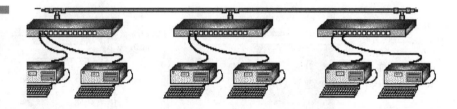

Figure 2-2
A configuration like this allows you to avoid excessive use of repeaters, and therefore also the propagation delay these can induce.

Daisy-Chaining, BNCs, and More. As we mentioned in the last section, 10 Base 2 segments can be interconnected by daisy-chaining them together. This extremely simple technique does have a drawback, though. It can create a network that is very vulnerable to problems caused by users' damaging or disconnecting cables. This is particularly true because a T adapter typically contains three BNC connections:

- a female connector to which a node's NIC plugs in
- two male connectors, each of which attaches to coax cable, as Fig. 2-3 illustrates

Figure 2-4 depicts network connections carried out with BNC/T adapters.

10 Base 2 Configuration Requirements and Bandwidth.
Repeaters, since they can greatly extend the reach of a 10 Base 2 network, may seem to be great pieces of equipment. But like everything, they have shortcomings, one of the most significant of which is that they

Figure 2-3
A T adapter must have two male BNC connectors, since no network device other than a terminator can be placed at the end of a cable segment.

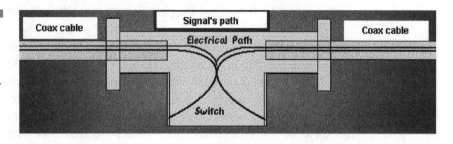

Coax cable

Signal's path

Electrical Path

Coax cable

Switch

Figure 2-4
As you can see, the single female BNC connector in a T adapter is needed too.

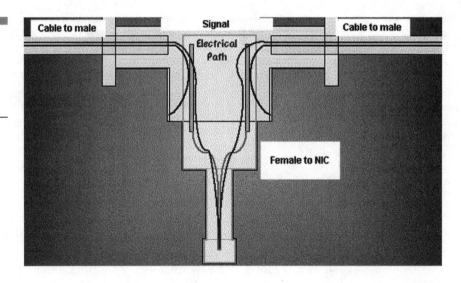

introduce a small amount of delay in the time it takes a signal to fully propagate through a network.

In order for CSMA/CD to work properly, it must detect collisions as they occur. It does so by monitoring data actually on the network, and comparing that data to what it's supposed to be transmitting. If CSMA/CD detects any difference, it assumes a collision has taken place, immediately stops sending, and then waits a random amount of time before transmitting again. This inherent CSMA/CD flaw, known as *propagation delay*, can be exacerbated by repeaters. What's more, if the propagation delay between a sending device and the device furthest from it on the network is larger than half the smallest frame size allowed, CSMA/CD can't properly detect collisions, causing data to be lost or garbled. So, 10 Base 2 and repeaters can introduce bandwidth overhead in more than one way.

To address such questions, the designers of Ethernet determined that in order to incur no more propagation delay than CSMA/CD could handle, a 10 Base 2 network should contain no more than four repeaters and five cable segments. These same calculations indicate that only three of these five segments could actually have devices attached to them. 10 Base 2's CSMA/CD-repeater constraints are therefore often called the *5-4-3 rule*. We've illustrated a network that conforms to this rule in Fig. 2-5.

Figure 2-5
Using 10 Base 2 to
cable your network
imposes limits on
more than segment
length.

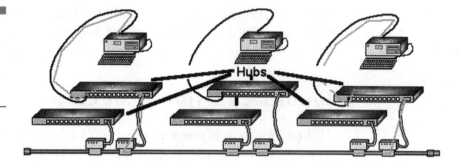

NOTE: *There's something very important to bear in mind when counting
repeaters. Only the number of repeaters between any two points on a network
need be taken into account. The total number of repeaters on the network as a
whole can be safely ignored.*

Look more closely at Fig. 2-5, and you should see that nodes A and D
are separated by three cable segments and two repeaters. Such a configuration, by staying below the limits imposed by the 5-4-3 rule, will present less bandwidth overhead than one that maxed out within that rule's
confines. Figure 2-6, on the other hand, illustrates a poorly designed net-

Figure 2-6
Here's a look at how
not to set up a 10
Base 2 network.

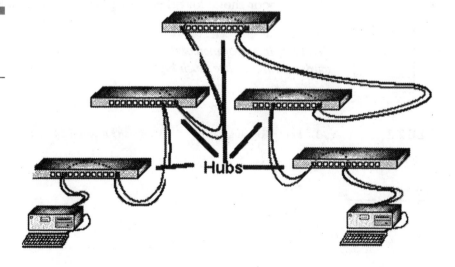

work that effectively ignores the 5-4-3 rule, and will probably therefore generate a very large number of collisions, unnecessarily consuming bandwidth as a result.

Unshielded Twisted Pair Cable

10 Base T or *twisted pair wire* actually consists of two insulated wires, usually 22- through 26-gauge copper wire, surrounded by a PVC or Teflon insulator, and twisted together in a spiral pattern. That twisting makes an important difference to the efficiency of 10 Base T, since it reduces low-frequency interference. Similarly, the common practice of using different twist lengths in adjacent pairs helps reduce the amount of crosstalk between pairs in a cable. This is particularly significant in those Category 3 or Ethernet 10 Base T implementations, because in these implementations, as many as several hundred such pairs can go into a single data communications channel.

10 Base T's major advantages are that it's inexpensive and easy to install. But don't run out or dial up to go cable shopping, because 10 Base T has a disadvantage that's particularly significant to anyone reading this book. Twisted pair offers limited bandwidth, particularly since transfer rates over the telephone lines to which it often connects can be as low as 300 to 9600 bps. You read that right—*bits*, not megabits, per second. While transfer rates up to 10 Mbps can be achieved for short, point-to-point UTP runs, the medium's inherent bandwidth busting makes 10 Base T a poor choice for other configurations, as Table 2-2 outlines.

NOTE: *In anticipation of the possibility that emerging data transfer rate standards, which mandate much higher speed cable, may require channels with*

TABLE 2-2

Summarizing 10 Base T's Applicability

In These Topologies	UDP Isn't a Good Choice
bus/backbone	X
tree	
ring	X
star	X

much improved bandwidth capabilities, many sites are installing Category 5
10 Base T, even though only Category 3 is actually required.

10 Base T Configuration Requirements and Bandwidth. As was
the case with 10 Base 2, 10 Base T limits network size in an effort to
control propagation delays. According to the IEEE requirements for
CSMA/CD, both types of cabling can support as many as 1024 nodes.
But UDP, unlike either Thinnet or Thicknet, carries with it no minimum distance between devices, since nodes that rely on 10 Base T can't
be connected serially anyway. The 10 Base T standards further require
that signal loss on a segment not exceed 11.5 dB, and state that the maximum amount of time between a collision being detected and a collision alert signal being placed on the network cannot exceed about 9
nanoseconds.

On 10 Base T networks, the MAU acts as repeater. The IEEE UDP standards address this dual role by requiring that propagation within an
MAU—that is, delay occurring between data arriving on the device and
being data passed on by it—must not exceed 2 nanoseconds.

There's more to consider. In UDP environments, the speed limitations
of the essentially dial-up medium are particularly apparent, and more a
factor even than in 10 Base 2 or 10 Base 5 environments, when *highly
interactive*, or *bandwidth-voracious*, applications are involved. Since highly
interactive and bandwidth-voracious describe much of what's currently
out there on the Internet, it behooves us to examine UDP's limitations
more closely. For 26-gauge copper wire such as that typically used in 10
Base T, attenuation is about 3 decibels per mile (3 dB/mi). The characteristic impedance of such wire is 671 ohms, while its raw velocity of signal
transfer is 19,000 miles per second.

We can use figures and formulas like those just cited to determine a
practical maximum segment length in 10 Base T networks, based on
more than one real-world, bandwidth-related shortcoming.

Since signal loss on a segment can't be more than 11.5 dB, and 26-
gauge UTP loses 3 dB per mile, the theoretical maximum length of a
UTP segment turns out to be:

$$\text{Maximum length} = 11.5 \text{ dB}/(3 \text{ dB/mile}) = 3.83 \text{ miles}$$

Similarly, any theoretical maximum length of segment that seeks to
minimize collision-generated bandwidth overhead must take into
account the following:

In a 10 Base T network, if A sends a packet that arrives at B just after B attempts to send a packet, a collision will be detected at B.

If the collision signal, sent by B as a result, arrives at A after A has started to send a second packet, A has no way of knowing whether the first or the current packet took part in the collision. Therefore, to be of any use, especially to our efforts to practice bandwidth frugality, a collision signal must arrive at A before A is done sending the packet involved in the collision.

Just to be on the safe side, we should assume that A deals in minimum-size packets of 512 bytes. So, we must set up network segments with a length that will permit a roundtrip transmit time no greater than 512 bits/10E + 06 Mbps, or 51.2 nanoseconds.

UTP Cable Attenuation Characteristics

We close this overview of UTP with a more in-depth look at its attenuation characteristics. Table 2-3 summarizes attenuation and its relationship to frequency for standard UTP cable consisting of 19, 22, 24, and 26 wires and adds, as a point of comparison, that same summary for UTP cable made up of three twists or braids of 24 wires each.

IN A NUTSHELL ▬▬ ▬▬ ▬▬ ▬▬ ▬▬ ▬▬

1. The two most commonly used forms of Ethernet cabling, copper coaxial and unshielded twisted pair, can each incur bandwidth overhead due to attenuation.

2. In addition to attenuation, coax can encounter bandwidth overhead, consisting of propagation delay engendered by repeaters.

3. Although 10 Base T networks don't experience repeater-generated propagation delay—because they cannot be daisy-chained—they can present other forms of bandwidth overhead, namely those due to:

   ▪ exceeding the 5-4-3 rule and thereby making collisions more likely

   ▪ configuring individual segments to a length that precludes collision messages being placed on the network in a timely fashion

TABLE 2-3	UTP Cable with This Many Wires	Transmitting at This Frequency	Will Incur Attenuation at This Rate
Summarizing Attenuation for 10 Base T	19	0.001 Mbps	0.746 decibels per kilometer (dB/km)
		0.01 Mbps	1.864 dB/km
		0.1 Mbps	3.846 dB/km
		1 Mbps	10.787 dB/km
		10 Mbps	31.895 dB/km
	22	0.001 Mbps	1.043 dB/km
		0.01 Mbps	3.183 dB/km
		0.1 Mbps	5.595 dB/km
		1 Mbps	14.638 dB/km
		10 Mbps	42.195 dB/km
	24	0.001 Mbps	1.346 dB/km
		0.01 Mbps	4.396 dB/km
		0.1 Mbps	7.298 dB/km
		1 Mbps	20.469 dB/km
		10 Mbps	59.593 dB/km
	26	0.001 Mbps	1.695 dB/km
		0.01 Mbps	5.591 dB/km
		0.1 Mbps	9.967 dB/km
		1 Mbps	27.494 dB/km
		10 Mbps	79.359 dB/km
	24 by 3	0.001 Mbps	0.256 dB/km
		0.01 Mbps	0.558 dB/km
		0.1 Mbps	0.984 dB/km
		1 Mbps	1.310 dB/km
		10 Mbps	1.752 dB/km

LOOKING AHEAD ▆ ▆ ▆ ▆ ▆ ▆

Chapter 3 picks up where this chapter leaves off, in investigating the effects of topologies in general and the spread or breadth of those topologies in particular on bandwidth usage.

Topologies and Bandwidth

This chapter examines the effects of the physical distribution and pattern of networks on bandwidth usage. We begin with a brief review of common topologies, or network layouts. Then we investigate the most widely used topologies in more detail.

Network Topologies

A network's *topology* is its physical layout or pattern. General topology types, such as bus, ring, and star, were designed based upon media access methods and cable types. What's more, these generalized network patterns can be varied and mixed-and-matched, so that, in large networks that span a wide physical area, you might find a grab bag of several topologies.

In this section, we scrutinize a number of categories of network topologies.

Bus or Backbone

In a *bus topology,* as Fig. 3-1 shows, every node, be it PC, printer, or other peripheral, attaches directly to a common cable. Needless to say, this method of connecting to a network would be inefficient in any but small environments. That's why a bus topology most often does not characterize networks as a whole, but rather only network *backbones,* to which either segments or clusters of devices then connect.

Figure 3-1
Take a look at a classic.

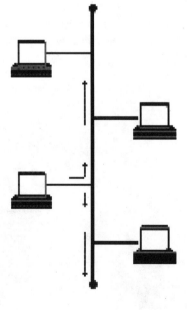

The wiring scheme of a classic bus topology lacks a central point of control, making this seemingly simple layout difficult to troubleshoot. When acting as a backbone around which workgroups or segments are clustered, this difficulty can be compounded.

Mesh

In a *mesh topology*, like that shown in Fig. 3-2, every station is connected to every other station.

Needless to say, a pattern like that shown in Fig. 3-2 is infrequently found in the real world. Mesh topologies' significance lies instead in their influencing other network designs.

Ring

Figure 3-3 illustrates a classic *ring topology*, which features a closed loop of cable. In a ring, messages travel in a single direction around the ring from one device to the next. Each network device acts as a repeater, regenerating signal as needed. Therefore, in a ring, if one node fails, the whole network can crash. This not-inconsequential disadvantage of the ring topology gave rise to a hybrid network pattern called the *star-wired ring*, which we discuss in a moment.

Figure 3-2

Meshes are not only far more expensive to set up, because of the additional connectivity hardware they require, but also much more subject to contention.

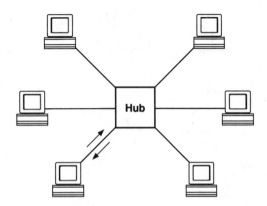

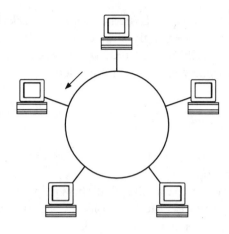

Figure 3-3
When healthy, rings provide excellent transmission speeds and optimal use of bandwidth.

Star

The *star topology* is, like the ring, one of the parents of the star-wired ring we examine next. Unlike a bus, a star is fairly easy to troubleshoot, because of its structured wiring scheme. Similarly, unlike the ring, one node's crashing won't create a domino effect in a star-based network.

In a star, as Fig. 3-4 depicts, every node has a dedicated set of wires connecting it to a central hub. Therefore, the failure of one node or connection seldom affects any others. At the same time and for the same reason—the presence of a hub—a star offers easily available network statistics, crucial to effecting not only optimum bandwidth usage but efficiency in a network as a whole.

Figure 3-4
The development of the star topology provided internetworking with an important tool—the hub.

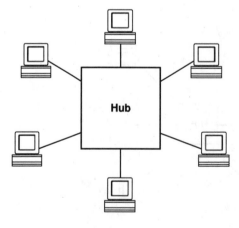

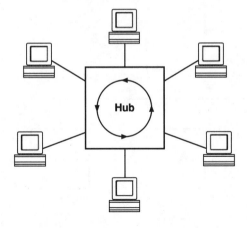

Figure 3-5
A star-wired ring can be considered the best of two worlds.

Star-Wired Ring

A *star-wired ring* like that in Fig. 3-5 has regularly replaced the ring topology in practical use. Networks based on star-wired ring topologies have nodes radiating from a wiring center or hub. The hub acts as a logical ring, within which data travels in a defined sequence from port to port. For this reason, the star-wired ring offers some degree of the lack of contention and efficiency of transmission that the classic ring does; but by incorporating the radiating nodes of a star, the star-wired ring avoids the classic ring's deadly tendency to fail at large if one node fails.

Tree

The *tree topology* can be considered a variant of both the mesh and the star. As Fig. 3-6 shows, a tree attempts to offer a high number of node-to-node paths, and therefore efficient transmission, while at the same time using hubs to minimize contention and provide a management window on critical network points.

Topologies and Their Effects on Bandwidth

Every layout we've discussed, and the real-world hybrids that draw on but at the same time vary these templates, have strengths and weaknesses, which we've categorized as:

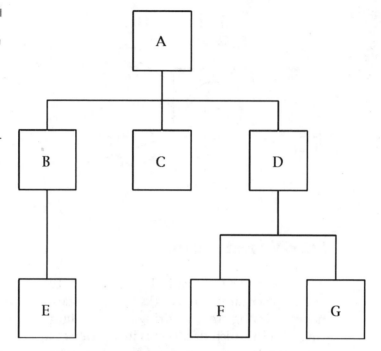

- contention, particularly as the number of stations increases
- ease of configuration and management
- effect on communications if a segment is damaged

Of these, the first clearly has the most impact on bandwidth.

Bus Topologies and Bandwidth

Bus topologies use CSMA/CD for coordinating transmissions. So, the way in which they handle collisions directly affects their bandwidth use.

As we discovered in Chap. 2, under CSMA/CD, a station won't begin transmitting until it senses an idle carrier. When collisions occur, that same station will once again wait for an idle line before it tries again. Therefore, under CSMA/CD on a bus, the key to efficient use of bandwidth is to provide each station with a varying but random amount of time to wait before resending. This can be accomplished by:

- assuming a fixed time slot of 512 bits or 51.2 microseconds at 10 Mbps

- permitting stations to begin transmitting only at the beginning of such a time slot

- after each collision, whose ordinal value we represent with the variable N, forcing a station to wait for an interval equal to either of $(N-1)$ or $(2 \cdot (N-1))$ time slots, selected randomly, before trying again.

NOTE: *This bus-based technique for avoiding collisions and therefore for improving bandwidth use is called binary exponential backoff.*

Ring Topologies and Bandwidth

Originally developed at IBM, *ring topology* (also frequently called *token ring topology* because it incurs no contention or collisions, and because the flow of traffic through it is strictly controlled) provides excellent transmission rates, usually in the range of 4 to 16 Mbps.

Collisions cannot occur on a ring because each node must take possession of a binary *token* (hence the colloquial name) or flag before it can transmit. In creating tokens, a practice called *bit-stuffing* is used to ensure that the pattern of bits in the token doesn't mimic any that might occur in data.

Once it's grabbed the token, a node on a ring network pushes data frames out onto the ring and then releases the token. Meanwhile, the transmitted data has begun to circulate through the ring, until it reaches its intended receiver.

While the highly structured nature of a token ring network provides clean transmission paths and efficient use of bandwidth, it can also contribute to poor performance on very dense or very busy networks, as Fig. 3-7 shows. While such performance losses aren't directly attributable to poor use of bandwidth as such, they certainly mimic it, at least from a user's point of view.

Star-Wired Ring Topologies and Bandwidth

While wired approximately like a star, the *star-wired ring*, because of its hub and the unidirectional flow of data within that hub, can:

Figure 3-7
When a token ring network becomes crowded or hyperactive, a station may have to wait quite a while to gain the token, and the ability to transmit.

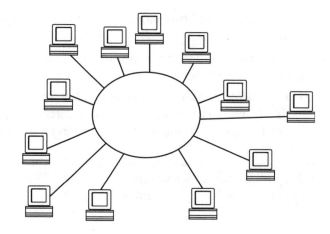

- preclude network downtime caused by failed cable segments; for instance, if a cable is cut, the hub detects this, and throws a bypass switch that isolates the relevant node, thereby allowing the remainder of the network to function

- provide more structured use of bandwidth than a star, which, like a bus, most frequently relies on CSMA/CD

- present less opportunity for lengthy waits to transmit than a ring

WANs

We've not talked about wide area networks so far in this chapter, only because, by their nature, WANs, which provide communications between dispersed geographic points, follow no specific topology. That's because WANs differ from LANs in a number of other significant ways, as Table 3-1 outlines.

WAN as Bottleneck

Since the Internet can be considered the quintessential WAN, working around the network bottlenecks—like that shown in Fig. 3-8, which can spring up at points of LAN/WAN interface—becomes critical.

Any device connecting LANs to WANs must control access to scarce WAN bandwidth. Among other things, such devices must:

- keep unnecessary traffic off the LAN/WAN connection

TABLE 3-1

Summarizing LAN/WAN Differences

LAN Characteristic	Handled on WANs
Switching as the primary means of directing traffic	By routing
A private cable plant as the primary transmission supply, with the resultant benefits:	Through the public telephone system, with the resultant disadvantages:
▪ plentiful bandwidth	▪ limitations on bandwidth
▪ relatively inexpensive bandwidth	▪ relatively expensive bandwidth
▪ good response time	▪ relatively poor response time

- ▪ reduce the amount of network overhead engendered by particular protocols
- ▪ provide some means of temporarily coming up with extra bandwidth when it's needed to overcome congestion

Routers, WANs, and Bandwidth

Routers, in addition to being the primary means of interface between LAN and WAN, also carry out one of the WAN bandwidth management tasks just cited. As Fig. 3-9 shows, routers can keep LAN traffic like:

Figure 3-8
A typical 64-Kbps WAN circuit provides $\frac{1}{160}$ the bandwidth of a 10 Mbps LAN.

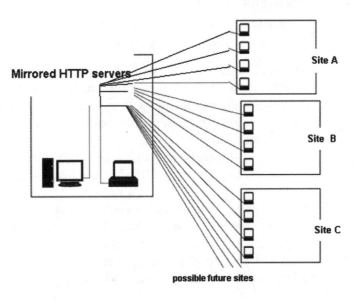

Mirrored HTTP servers

Site A

Site B

Site C

possible future sites

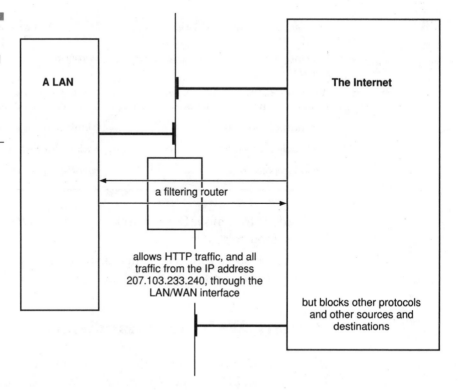

Figure 3-9
By filtering unsupported protocols and traffic from or to unrecognized sources, routers in effect also serve as firewalls of sorts.

- broadcast traffic
- traffic from unsupported protocols
- traffic destined for unknown networks

from crossing the link to the WAN.

Routers examine every packet they receive, which makes them an excellent choice for controlling, queuing, and prioritizing traffic across the LAN/WAN interface. However, their message-managing skill is at the same time the very thing that can introduce, rather than remove, overhead from LAN-to-Internet communications. To begin to understand why this is so, let's investigate the LAN/WAN interface a little more closely.

WAN Routing Protocols. Two broad categories of routing protocols exist: distance-vector and link-state protocols. As Fig. 3-10 depicts, the *distance-vector routing protocols* on a given router will periodically transmit the router's entire routing table to each of the device's neighbors, even if changes in network status haven't taken place.

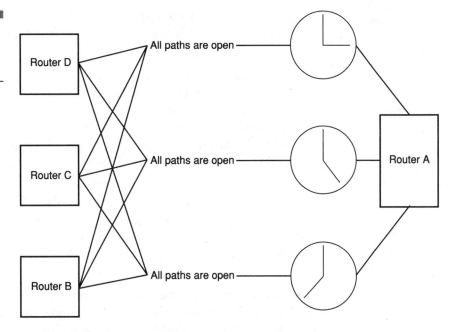

Figure 3-10
The latter two transmissions shown here were unnecessary.

Link-state routing protocols, on the other hand, transmit smaller update packets, usually only when an actual change in network conditions has taken place. As a result, link-state routing protocols, although they consume more severe CPU time than their distance-vector siblings, consume much less bandwidth.

Link-state protocols outperform distance-vector methods in other ways. When a network link fails, link-state protocols recalculate routes and therefore can forward traffic much more efficiently. In addition, link-state protocols support hierarchical routing, which permit a LAN to be divided into several individual routing domains, thereby giving individual routers less to handle.

Dial-up Access and the LAN/WAN Connection. When communicating between routers across a WAN, probably the best way to jazz up the ordinarily stodgy nature of dial-up communications is to use a link operating in dialup-on-demand or DOD mode. Such connections can be brought up and down depending on the traffic on a given link. What's more, with DOD, if a primary WAN connection begins to experience congestion because of increased traffic, some routers can automatically activate a second-string dial-up line to provide additional bandwidth.

Using Compression to Speed Up WANs.   One obvious way to squeeze more bandwidth out of a LAN/WAN link is to use data compression. Two basic types are available:

1. *History-based compression* looks for repetitive data patterns across multiple packets and replaces them with shorter codes. Then, both sender and receiver build a dictionary and encode and decode transmissions according to that dictionary. Since history information, that is, the repetitive patterns in data, are transferred right alongside compressed data itself, this form of compression can operate well only over reliable links.

2. *Per-packet compression,* on the other hand, doesn't need to concern itself with the reliability of paths, because it uses no history information. It looks for repetitive patterns within every datagram and replaces these patterns with shorter codes.

Using Bandwidth Aggregation to Speed Up WANs.   By its nature, data communications traffic is, in the jargon of the field, *bursty.* To understand what this means, consider the following scenario:

- A network is quiet for a brief period.
- Then the network experiences a quick burst of activity, in the form of the receipt by one of its clients of a Web page downloaded from an Internet site.
- The network becomes quiet again for a while, and then...

Clearly, such unpredictability is a challenge to LAN/WAN connections. If such an interface is too frequency generous, bandwidth that may never be used will still have to be paid for. If the connection is too stingy, the link may experience:

- long transmission queues
- as a result, unnecessarily retransmitted frames, dropped packets, slow response times, and session timeouts

One technique, called *bandwidth aggregation,* which attempts to circumvent such problems by providing flexibility in defining LAN/WAN link speeds, relies on the *multilink point-to-point protocol,* or MPPP. This protocol is in effect an algorithm for splitting, recombining, and sequencing datagrams across multiple transmission paths. The designers of MPPP were originally motivated by the desire to exploit multiple bearer channels such as those offered by ISDN. But the protocol has been developed to be just as applicable to any situation in which multiple PPP links connect two systems. Bandwidth aggregation even includes thresholds that can be configured to allow the size of the LAN/WAN connection to vary with load.

Prioritizing Data to Speed Up Access to WANs. Under high-traffic conditions, LAN/WAN interfaces often must combine time-sensitive traffic such as that generated by a terminal session with batch traffic like that pertaining to a file download. Efficiently sharing the limited amount of available bandwidth while still meeting the specific needs of different types of transmissions often requires *prioritization*. This technique offers the ability to distinguish between time-sensitive and batch traffic, and to give the former higher priority in the WAN transmission queue. Typically, prioritization assigns a network-manager-defined priority to every packet that's getting ready to leave the LAN for the WAN, and then forwards each packet to queues designated as:

- urgent priority
- high priority
- medium priority
- low priority

The router that acts as interface then forwards all packets in the urgent-priority queue, then those in the high-priority, and so on.

Reserving Protocols to Speed Up WANs. Some applications require a specific percentage or range of frequencies of bandwidth. The technique known as *protocol reservation* allows administrators to reserve a portion of a LAN/WAN interface's bandwidth for a specific protocol or application. Protocol reservation particularly benefits interactive applications, by configuring specific percentages of a link's bandwidth for every protocol that operates over the LAN/WAN link. So, for instance, if 27 percent of an interface's bandwidth has been reserved for FTP traffic, FTP sessions will always have that amount of bandwidth available, whatever the traffic levels on the remainder of the interface's bandwidth pipeline, unless FTP-generated traffic consistently drops below its reserved 27 percent. Should that happen, leftover bandwidth may be offered to other protocols and applications.

Using Session Fairness to Speed Up Access to WANs.   *Session fairness,* an enhancement to protocol reservation, seeks to ensure that traffic from all users is forwarded at an equal pace, so that no single user can monopolize the LAN/WAN connection. Session fairness, when used in combination with protocol reservation, can be a powerful tool. Let's sketch an example.

Suppose the exponentially growing, brilliantly successful (but fictitious) Web page design firm of Desktop Data decides to increase its Web presence. By definition, the traffic generated in this environment would consist largely of HTTP packets. So, configuring protocol reservation is a relative no-brainer. But, if session fairness is overlooked, HTTP traffic from the marketing department might dominate the entire range of bandwidth reserved for HTTP, leaving such departments as Technical Support and production in the lurch. However, once session fairness is incorporated into the configuration of Desktop Data's Web server, the 65 percent of that server's LAN/WAN interface bandwidth that has been set aside for HTTP traffic will be shared equally among all concurrent HTTP sessions, however many there might be.

Ranking Packets to Speed Up Access to WANs.   *Packet-ranking* builds on both data prioritization and protocol reservation. Such ranking lets network administrators specifically identify particularly important messages, down to the level of individual packets. By so doing, packet-ranking can:

- move traffic through a router's internal buffers more quickly
- dynamically rearrange the router's queues
- reduce delays in delivery

WANs and Multicasting.   Already an Excedrin headache level three for many network managers, and getting more severe all the time, is the riddle of how to efficiently deliver multimedia content to end users across the Internet. This one's a particular puzzler because the same information may have to be distributed to many different sites, meaning that the same huge audio, graphics, and video files must travel the same overworked LAN/WAN interface time and time again.

The technique known as *multicasting* seeks to put a stop to such obvious bandwidth frittering by sending out files only once, and then replicating the files at specific points in their delivery path. Multicasting relies on routing protocols that create a shortest-path distribution tree from the source to all destinations. The source, of course, occupies the root of such a tree; packets flow from this root across the tree's branches, directed by controlling messages which the sender injects into the data stream, and which take care of such details as determining if a particular branch has no currently logged-in users.

Multicasting can reduce bandwidth consumption across LAN/WAN interfaces by ensuring that:

- ■ a sender transmits only one copy of a body of data
- ■ a message takes the shortest path available to it across the distribution tree to the receiving node
- ■ a message crosses the LAN/WAN interface only once
- ■ a message is replicated and forwarded to specific areas of the distribution tree only if there's someone there to pick it up

IN A NUTSHELL

1. The bus topology can create bandwidth overhead through its handling of collisions.
2. The ring topology, although ordinarily performing well, can experience delays under high-traffic conditions.
3. A number of hybrid topologies such as the star-wired ring seek to blend the best characteristics of their source topologies, while avoiding those sources' shortcomings.
4. LAN-to-WAN connections offer their own collection of bandwidth problems, since the interface between the two types of networks can operate no faster than the WAN itself.

LOOKING AHEAD

Having inspected the effects on bandwidth of topologies, we turn in Chap. 4 to investigating the effects of one corollary of topologies: connectivity methods.

CHAPTER 4

Connectivity Methods and Demands on Bandwidth

In this chapter, we scrutinize the three connectivity methods upon which the bulk of Internet, or Intranet-to-Internet, traffic rely:

- dial-up
- leased line
- Integrated Digital Service Network, or ISDN

In each case, we pay particular attention to data transfer rates and bandwidth spectra.

Dial-up

Dial-up connections—although they have a number of advantages, such as being relatively inexpensive and of course universally available—are the least efficient connectivity method in terms of efficiency of bandwidth. Dial-up connections rely on asynchronous paths and serial connections to those paths. What's more, in the United States, the wiring that characterizes the telephone system, that is, the single pair of copper wires that makes up UTP, can at best transfer data at about 56,000 bps. When called upon to move data of a variety of formats simultaneously, as must be done, for example, in real-time transfers of multimedia material, the physical limitations inherent in dial-up connections cause them to fall even further short of meeting user demands. Put simply, when you have only one transmission channel to work with, there's no way to divide up the bandwidth spectrum. So there's no way to distribute, concurrently and in a coordinated way, data streams of a number of different make-ups. Nor is there any way to accomplish another requirement of real-time multimedia data transfer—streaming such data continuously, forgoing start, stop, and parity markers.

Leased Lines

Leased lines are just what their name suggests, the reserving of an actual wire or part of a physical channel for your own use. Leased lines come in a number of varieties; two of the most common are, to use AT&T's nomenclature:

- T1, like those shown in Fig. 4-1, running at speeds up to 1,544,000 bps or 1.544 Mbps
- T3, depicted in Fig. 4-2, transferring at up to 45 Mbps

There's nothing magical or particularly cutting-edge about either T1 or T3. Both are amalgams, the former being a collection of as many as 24 standard UDP pairs, and the latter an aggregation of as many as 24 T1 lines.

Given this conglomerate nature, it's easy to understand why T1 or T3 offers not only data transfer rates that are far superior to that of dial-up but also a much greater available bandwidth spectrum—more physical

Figure 4-1
A T1 line consists of up to two dozen UTP pairs.

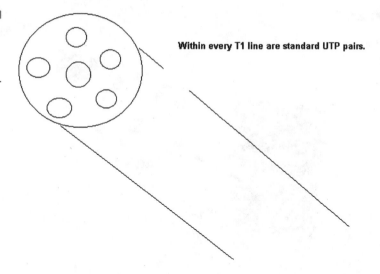

Within every T1 line are standard UTP pairs.

paths exist. These additional pathways make it possible, for instance, to stream a variety of data simultaneously and efficiently.

But leased lines, although presenting you with a much wider range of available bandwidth, do have drawbacks. The lines themselves, and the hardware needed to connect to them, are much more costly than those

Figure 4-2
T3 bundles T1 lines rather than standard UTP.

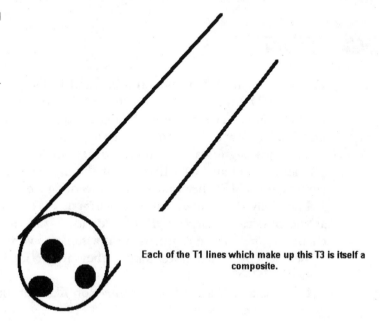

Each of the T1 lines which make up this T3 is itself a composite.

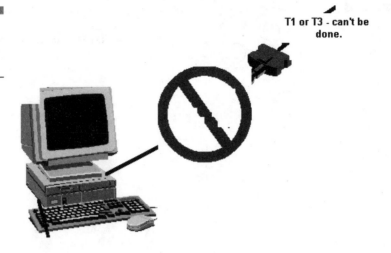

T1 or T3 - can't be
done.

required by dial-up connections. What's more, in a sort of bandwidth-Catch-22, leased lines are sometimes *too* fast, in the sense that a PC's serial port can't keep up with the transfer rates typical of either T1 or T3, as Fig. 4-3 attempts to illustrate.

In other words, a PC can't be directly connected to a leased line.

ISDN

Bell Atlantic, in its literature on residential ISDN service, emphasizes the fact that ISDN transfers data four to five times as fast as analog modems, topping out at 128 Kbps. Since ISDN, like dial-up, relies on UTP, albeit a bunch of it, how is this accomplished?

The most significant additional difference between a standard telephone line and an ISDN line is that the former is analog, whereas the latter is digital. In other words, ISDN was designed to support data.

Effectively, the *plain old telephone system* or POTS wiring, when serving as the physical carrier for ISDN, is divided into three separate logical channels. These three channels are not separate wires or wire pairs, like those used by leased lines. Rather, they are defined by the logic that underlies ISDN.

Let's use a residential ISDN connection like that illustrated in Fig. 4-4 as an example.

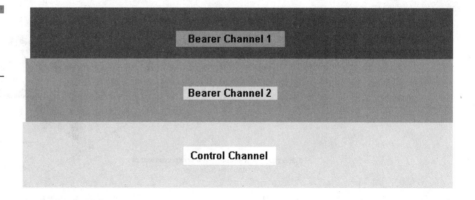

Figure 4-4
ISDN's transmission channels are logical, not physical.

Three channels exist in such a connection. Two are 64 Kbps *Bearer* or B channels; the other is a 16 Kbps *Delta* (D) channel. (This configuration, common to many types of ISDN service, is often called 2B + D) The two B channels of such an ISDN connection, also known as *basic rate interface* or BRI, deliver the bulk of a transmission, be it voice or user data. Configured to carry only the latter, a single standard ISDN connection can indeed transmit at 128 Kbps. BRI's D channel carries only control data. Whatever the voice and/or data transfer capacity of an ISDN connection, only one such channel must be present. So, for example, if you were to sign up for multiple ISDN lines multiplexed together, you would still need only one D band. Even residential ISDN connections can now take advantage of such multiple and multiplexed lines, like those offered by Bell Atlantic and other carriers, called *primary rate interface* or PRI connections. PRI consists of 12 B and 1 D channels, and offers transfer rates between 760 Kbps and 1.544 Mbps.

ISDN Pros and Cons

Let's tick off the most significant benefits ISDN offers.

1. ISDN, by virtue of its multiple logical channels, can simultaneously move many different types of information, including voice, text, images, audio, and video.

2. ISDN's data transfer rates involve no compression. When compression is added, a standard BRI connection can move data at up to 512 Kbps.

3. Most frequently, ISDN can use existing POTS wiring.

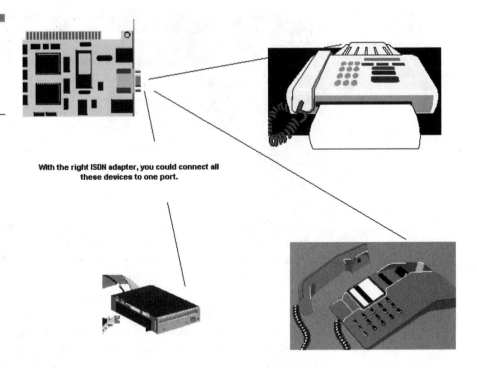

Figure 4-5
ISDN can act as a data transfer analog to small computer systems interface (SCSI) connectors.

With the right ISDN adapter, you could connect all
these devices to one port.

4. As Fig. 4-5 shows, ISDN can support as many as eight devices on a single line, allowing you, for example, to use one ISDN connection for:

- voice and data
- a modem and a modem server
- voice, a FAX server, and a modem server
- voice, a FAX server, a modem server, and a printer

5. ISDN is an on-demand service, and therefore less expensive than dedicated leased lines.

Now the downside.

1. Just like leased lines, ISDN connections need special hardware, which can be quite costly. For instance, while preparing this chapter we skimmed a number of computer products catalogs and Web sites in order to compare the prices of 56 Kbps modems and ISDN adapters. The latter proved to cost roughly twice as much as the former.

2. Not all ISDN adapters and routers provide standard analog connections, which means that such connection-challenged devices may deny you the use of some analog equipment like FAXes.

Comparing Connectivity Methods

We close this chapter with a comparison, presented in Table 4-1, of the relative data transfer performance of dial-up, ISDN, and leased-line connections.

TABLE 4-1

Dial-up, ISDN, and Leased-Line Transfer Rates and Bandwidth

Connection	Items Transferred	Time
9.6 Kbps analog	Screen (50 KB)	42 seconds
	Image file (1 MB)	14 minutes
	Multimedia (50 MB)	11.8 hours
14.4 Kbps analog	Screen (50 KB)	27 seconds
	Image file (1 MB)	9 minutes
	Multimedia (50 MB)	7.7 hours
28.8 Kbps analog	Screen (50 KB)	13 seconds
	Image file (1 MB)	5 minutes
	Multimedia (50 MB)	3.8 hours
56.6 Kbps analog	Screen (50 KB)	7 seconds
	Image file (1 MB)	3 minutes
	Multimedia (50 MB)	2 hours
64 Kbps ISDN	Screen (50 KB)	6 seconds
	Image file (1 MB)	2 minutes
	Multimedia (50 MB)	1.7 hours
128 Kbps ISDN	Screen (50 KB)	3 seconds
	Image file (1 MB)	62 seconds
	Multimedia (50 MB)	52 minutes
1.54 Mbps PRI or T1	Screen (50 KB)	3 seconds
	Image file (1 MB)	5 seconds
	Multimedia (50 MB)	4 minutes

IN A NUTSHELL

1. Standard UTP-based dial-up connections cannot transfer at rates faster than 56 Kbps.
2. Dial-up connections are inadequate for the transfer of multimedia material.
3. Leased lines, both T1 and T3, consist of bundles of lesser-capacity channels.
4. Leased lines cannot be directly connected to analog modems.
5. ISDN offers data transfer rates comparable to those of leased lines.
6. ISDN hardware, like that used with leased lines, is more expensive than standard analog connection hardware.

LOOKING AHEAD

In Chap. 5, we examine the effects on bandwidth of connectivity hardware small and large: modems, NICs, routers, bridges, and hubs.

CHAPTER 5

Connectivity Hardware and Demands on Bandwidth

In this chapter, we examine a spectrum of connectivity hardware, from the tiniest (the UAR/T chip) to the beefiest (bridges, routers, hubs, and switches). Specifically, we investigate the inner workings of:

- RS-232 serial communications, the UAR/T chip, and the timing of transmissions and receptions
- NICs and their packet drivers
- bridges
- routers
- hubs

in terms of their effect on bandwidth consumption.

Serial (RS-232) Communications

If you connect to the Internet by anything other than a dedicated leased line, you're connecting through a modem of some sort. So the first thing we scrutinize in this chapter is modem-based transmissions.

The serial port to which a modem attaches operates under specific design constraints. As Fig. 5-1 indicates, in *RS-232 communications,* a pulse of from −3 to −15 volts represents a *logical 1,* while a pulse of +3 to +15 volts represents a *logical 0.* A serial line's default condition is ordinarily logical 1.

Before it does anything else, asynchronous data transmission sends a start bit, in order to inform the receiver that the logical 1 it senses is the start of a message, and not a data bit. After this start bit, a number of data bits, whose specific value depends on the encoding method employed, travel out. Then we begin to see what can be considered over-head in serial communications: the optional parity at the end of a message, which attempts to provide error detection. After the parity bit, one or two stop bits will be present, in order to signal the receiver that it has gotten one character. So, assuming that we're dealing with an ASCII-based transmission, and assuming further that the message in question uses one parity bit, the ratio of data bits to control bits is only slightly more than 2 to 1. As Fig. 5-2 illustrates, for every character we ship out under RS-232 serial communications, we must provide, not the 7 bits needed to encode that character, but 10.

Figure 5-1

Serial communications rely on two signal states, but a spectrum of voltages.

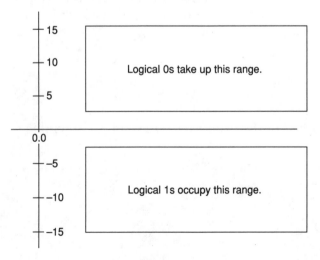

15
10
5
0.0
−5
−10
−15

Logical 0s take up this range.

Logical 1s occupy this range.

Figure 5-2
Particularly for ASCII-encoded transmissions, serial communications introduce significant overhead from the start.

This single ASCII character

must make room for much more than data

| start bit | data | parity bit | stop bit |

UAR/T as Receiver

As important to the performance of serial communications as its encoding patterns is its transmission mechanism, the *universal asynchronous receiver/transmitter* (UAR/T) chip. In a UAR/T chip, a receiver area detects a start bit by sampling each bit it senses, looking for a logical 0. When it thinks it has found one, it assumes it has found a start bit. But the chip wants to be sure. So, as you can see in Fig. 5-3, a mechanism within the chip called a *programmable divider* counts off enough receiver clock pulses to allow it to reach the center of this first received pulse. Then, once the programmable counter has arrived there, a sample is taken of the bit that

Figure 5-3
UAR/T timing is more involved than you might think.

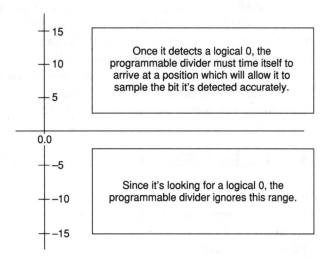

15

10

5

Once it detects a logical 0, the programmable divider must time itself to arrive at a position which will allow it to sample the bit it's detected accurately.

0.0

−5

−10

Since it's looking for a logical 0, the programmable divider ignores this range.

−15

the UAR/T chip uses to verify that the pulse it has encountered is indeed a start bit, not simply transient noise.

There's more to how the UAR/T's programmable divider operates, but we think you get the idea. Even at this minuscule level, the operations required to make sense of data transmissions are nearly as bandwidth-consuming as the transmissions themselves. And there are more such operations, for example, the techniques applied in an effort to prevent timing errors.

Timing Errors

If an RS-232 receiver clock is even (you should pardon the pun) a bit slower than its transmitter clock, pulse length as counted by the receiver's programmable divider will not be the same length as that sent out by the transmitter. Therefore, despite its assiduous sampling of the data stream, the receiver may, for want of a better metaphor, conceive of that stream's bits as being either slightly less or slightly more than the required distance—or more correctly, time interval—apart. Such lack of coordination can cause transmission errors due to errors in sampling.

Timing Requirements

Various encoding methods have various UAR/T timing requirements. In the outline of those requirements given in Table 5-1, we've used P to indicate actual pulse duration and D to denote pulse duration as counted by the receiver's programmable divider. What we've come up with is the range of timing within which UAR/T pairs must operate in order to remain adequately synchronized and therefore to preclude unnecessary bandwidth consumption caused by errors in transmission.

UAR/T Clock Performance

Now let's examine the actual performance of three typical UAR/T chips. The numbers presented in Table 5-2 were obtained from the chips' manufacturers.

Table 5-2's message is clear. Recent UAR/Ts from top manufacturers

TABLE 5-1

UAR/T Clock Frequency Ranges

Encoding Method	Encoding Pattern	Receiver Slow?	Receiver Fast?	Required Receiver UAR/T Clock Frequency
Binary Coded Decimal (BCD)	1 start bit, 6 data bits, no parity bit, 1 stop bit	7.5D must be greater than 7P, or 7.5D > 7P, as in D/P > .933. That is, the receiver UAR/T clock must be no more than 6.67% slower than the transmitter clock.	7.5D must be less than 8P, or 7.5D < 8P, as in D/P < 1.0667. That is, the receiver UAR/T clock must be no more than 6.67% faster than the transmitter clock.	Must be within ±6.67% of the transmitter
ASCII	7 data bits, 1 parity bit 1 stop bit	9.5D must be greater than 9P, or 9.5D > 9P, as in D/P > .947. That is, the receiver clock must be no more than 5.26% slower than the transmitter.	9.5D must be less than 10P, or 9.5D < 10P, as in D/P < 1.052 That is, the receiver clock must be no more than 5.26% faster than the transmitter.	Must be within ±5.26% of the transmitter
Extended Binary Coded Decimal Interchange Code (EBCDIC)	8 data bits, 1 parity bit, 1 stop bit	10.5D must be greater than 10P, or 10.5D > 10P, as in D/P > .952. That is, the receiver clock must be no more than 4.76% slower than the transmitter.	10.5D must be less than 11P, or 10.5D < 11P, as in D/P < 1.0476. That is, the receiver clock must be no more than 4.76% faster than the transmitter.	Must be within ±4.76% of the transmitter

TABLE 5-2

UAR/T Performance

Chip	Manufacturer	Largest Error Rate (at 56,000 bps)	Meets Requirements of Table 5-1?
PC16550D	National Semiconductor	±2.86%	Yes
PC16552D Dual	National Semiconductor	±2.86%	Yes
TL16C550A	Texas Instruments	±2.86%	Yes

have been engineered to avoid, as much as possible, unnecessary bandwidth consumption.

NICs and Packet Drivers

Within a LAN, the *network interface card* or NIC plays as essential a role as a modem does between LANs. In this section, we scrutinize these basic devices.

Like printers prior to the introduction of the PostScript and HP protocols, NICs, varying widely from manufacturer to manufacturer, originally had no standard software available. Then, the FTP Corporation defined a standard software interface for NICs running on PCs. This interface allowed NIC manufacturers to do no more than supply a small *terminate-and-stay-resident* (TSR) program, called a *packet driver.*

Packet drivers provide a common programming interface that allows a variety of protocol suites to use the same network adapter. Without these drivers, network software would have to be customized for every adapter type, and mixing of protocols would be virtually impossible. The original packet driver specification describes the programming interface to the FTP protocol; almost all packet drivers conform to this standard.

Three levels of packet drivers are described in the FTP spec: basic, extended, and high-performance. The extended package supports multicasting and statistics gathering, whereas the high-performance package offers performance improvements and tuning. Since such administrative tasks clearly involve a modicum of bandwidth overhead, we confine the rest of our discussion of packet drivers to the basic category. We think you'll agree that the techniques employed by these drivers must be taken into account in understanding bandwidth usage.

Identifying Network Types

NIC packet drivers perform essential identification tasks as part of their job of providing support for a number of protocol and hardware environments. Such identification relies on a group of integers, each of which represents parameters such as media type (e.g., Ethernet, Token Ring, and so on) the NIC can handle.

Driver Operations

A packet driver begins operating by means of a software interrupt in the range of 60 to 80 hexadecimal. Then, if the NIC is to support multiple protocols, the operating system must determine which protocol a packet is intended for. This latter task most frequently resides with some part of a message header. For instance, in standard Ethernet transmissions, such decisions would be based on the contents of a 16-bit *ethertype* field, which immediately follows destination and source addresses in a message header.

Next, a NIC, through its packet driver, must call a low-level function in order to define a destination for the type of link-layer packet that's been determined to be needed. The packet driver uses internal registers to store information about both NIC and the type of packets it handles. Function calls like these, defined in the FTP spec, can be programmed in either C or assembly language.

Packet Driver Function Calls

Since they're critical to a NIC's functioning, we investigate packet driver functions more closely, after outlining their interactions in Fig. 5-4.

`driver_info()` This function returns information about the NIC itself, and can supply:

- an internal hardware identifier
- the class of medium and type of card

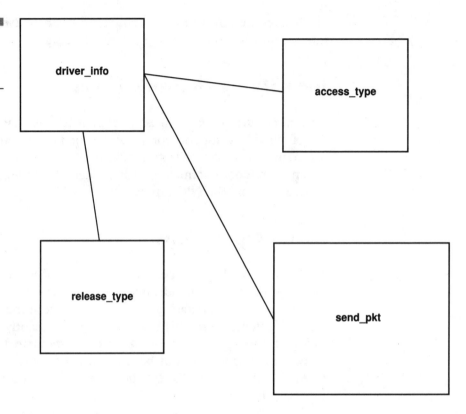

Figure 5-4
Standardizing NICs
requires significant
software interaction.

- the number of cards of a given class and type that are present
- the level of sophistication of the driver, that is, whether it operates in basic, extended, or high-performance mode

access_type( )   This function is the module that actually kick-starts access to packets of the appropriate type; access_type needs three arguments:

- type, which it gets from the function driver_info
- typelen, an integer which represents the number of bytes in, that is, the length of, the type field
- a pointer to another module, which will be called when a packet is received

When a packet is received, a first call to the receiver module asks for a buffer into which it may copy the packet. Then, a second call is made to this module, in order to verify that the packet has indeed been copied to the buffer.

release_type( )   This function is the packet driver's analog to closing a file; release_type terminates access to packets of a particular type.

send_pkt( )   This function takes two arguments:

- length, representing the number of data bytes to be sent
- buffer, a pointer to the buffer into which packets will be received

It's worth pointing out here that any application calling a packet driver must supply a complete packet, including local network headers, if send_pkt is to be able to function properly.

terminate( )   This function causes the driver being used to exit, and thereby allows the operating system to reclaim the memory that driver had been taking up.

get_address( )   This module writes the current LAN address of the NIC into a specific buffer, and returns an integer indicating the actual number of bytes written into a specific register.

reset_interface( )   This module is called when, for whatever reason, a transmission must be aborted. In essence, reset_interface reinitializes a NIC.

Suffice it to say that the efficiency with which the software modules that make up NIC packet drivers interact can affect the efficiency with which those drivers and NICs manipulate available bandwidth.

Bridges

Bridges allow networks to be extended beyond their default cable run lengths, and even past the segment lengths repeaters might provide. Bridges do this by capturing and then retransmitting packets. Such tasks, of course, require the bridge to store a number of parameters, most important among them MAC-level addresses. In this section, we scrutinize the efficiency of a number of bridges by examining their address tables.

Table 5-3

Bridge Efficiency as
a Function of
Addressing

Manufacturer	Model	Number of MAC Addresses Stored	Size of the MAC Address Table
RAD (Northwest Technical Services)	TrimBridge-10	4096	24 KB
Accton	Multi-Port Remote Ethernet Bridge-iR	8192	48 KB
Ragula Systems	100 VG AnyLAN	256	1.5 KB

All the bridges we look at meet the following criteria:

- they can connect Ethernet segments
- they can direct TCP/IP packets to a destination

Table 5-3 outlines the bridges we're examining.

To better understand the role of the MAC address table in bandwidth consumption, let's examine the environment in which the bridges described in Table 5-3, and many others, were tested—Worcester Polytechnic Institute. All these bridges draw upon two files that contain configuration information for the WPI network: (1) *hosts,* which holds IP addresses for all machines on the network and (2) *ethers,* which contains the MAC addresses for these computers. At the time this book was being written, the hosts file recorded the IP addresses of 2196 computers. Therefore, a bridge like that which Fig. 5-5 depicts, in order to serve this network at all, would have to have the ability to store at least this many MAC addresses.

Figure 5-5

In heterogeneous
protocol environ-
ments, a bridge must
also store router
addresses.

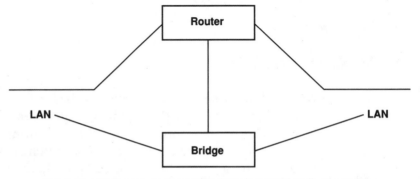

This bridge filters packets, forwarding those it cannot handle to a router.

In order to serve the network efficiently, that is, in order to be able to maximize the speed at which it can forward packets, such a bridge should be able not only to store all current addresses but also to provide for future growth of the network. Add to this the fact that Ethernet bridges build their MAC address tables themselves, by monitoring network traffic and packet sources. Therefore, such tables are to some degree volatile, making a bridge with the ability to store only a limited number of MAC addresses even less appropriate.

Routers

Like bridges, *routers* function to extend the expanse of networks. Unlike either bridges or repeaters, though, routers deal in higher-level addressing, and thereby can connect networks or network segments that do not share protocol environments. Still, the role of routers in bandwidth consumption can be measured with a yardstick very much like that we applied to bridges. In this section, we investigate five commercially available routers and the size of their routing tables, using Table 5-4 to accomplish that.

Routers, unlike bridges, often must carry out protocol conversion and other similarly sophisticated tasks. Therefore, routers cannot be evaluated primarily on the basis of their addressing efficiency. To delve deeper into what makes routers tick and what gives them an effect on bandwidth consumption, let's examine one of the routers from Table 5-4 more closely.

TABLE 5-4

Summarizing Routing Tables

Manufacturer	Model	Stores This Many IP Addresses
Xyplex	511/512 Ethernet-To-FDDI Router	32,000
Retix	RouterXchange 7500/7550	16,000
SVEC	FD2600H BRouter/Smart Master EtherHub	8,192
APT Communications	ComTalk HX Router	4,000
ConnectWare	EtherConnect Local Router	2,100

The FD2600H BRouterSmart Master EtherHub integrates a remote bridge, router, and SNMP master hub in one unit. Therefore, it supports wide area remote bridging, frontier routing, SNMP management, and Ethernet repeater functions, as well as intersegment or inter-LAN routing. What's more, the FD2600H can manage up to four other compatible hubs, thereby allowing all five devices to occupy just one IP address. Such combined addressing would in turn serve to streamline routing tables and the use of routing protocols in any network that contained such a device.

The Smart Master EtherHub's remote bridging relies on the IEEE 802.1d Spanning Tree Protocol, which in effect supplies an automatic learning algorithm to allow the device to build up its address tables in the most efficient manner possible. By virtue of this algorithm, the Smart Master acts as a central router to the LAN nodes it serves, while at the same time performing to some extent as an inter-LAN router, by forwarding all nonlocal packets to the transmission path best suited to them. The FD2600H supports multiple protocols within LANs, and uses the Internet *routing information protocol* or RIP to route IP packets to a WAN link, as well as a similar protocol from Novell to do the same for IPX packets. This device complies with the IEEE 802.3, 10 BASE 5, 10 BASE 2, and 10 BASE T standards. On any of those media, it can carry out packet filtering at rates as high as 14,880 packets per second (pps) for LAN ports it serves. On leased lines, its performance is, as you might expect, even better. Across such media, the FD2600H forwards data at a rate of from 1.544 to 2.048 Mbps.

The point is this. While the Smart Master and similar devices may not lead the pack in address storage capabilities, other characteristics, such as filtering and data transfer rates, can contribute to efficiency of bandwidth use as great as devices with larger routing tables.

Hubs

We have one more category of connectivity device to examine, and therefore (you guessed it) one more table to scan. Table 5-5 summarizes the makeup of commercially available hubs that are widely used on the Internet.

Of the hubs investigated in the study from which we drew Table 5-5, Cisco's Catalyst 5000 was chosen the overall winner, despite its ranking only

TABLE 5-5

Capabilities of
Common Internet
Hubs

Vendor	Model	Ports Supplied	IP Addresses per Port	Total Addresses Offered
CrossComm Corp.	Ethernet Workgroup Switching Hub	256	32,000	8,192,000
3Com Corp.	LANplex 6004 Switching Hub	48	8,192	393,216
Cisco Systems Inc.	Catalyst 5000	96	1,333	127,968
Fibronics International Inc.	Gigahub	100	1,024	102,400
Bay Networks Inc.	Lattisswitch Model 28115	16	1,024	16,384
Digital Equipment Corporation	DECswitch 900EF	12	1,333	15,996

third in the number of IP addresses offered. This device was, for instance, subjected to a 200 percent overload on all ports for three minutes. During this test, it did not lose so much as one packet. The Catalyst 5000, like the FD2600H router, demonstrates that sheer addressing capacity is not the only criterion by which connectivity hardware should be judged.

IN A NUTSHELL ▬▬ ▬ ▬ ▬ ▬ ▬

1. The synchronization required by sending and receiving modems adds bandwidth overhead to any environment.

2. The software modules used by NIC packet drivers, although they do not directly draw upon bandwidth, do affect the speeds at which data communications take place within LANs.

3. The number of MAC addresses bridges can store affects their use of bandwidth by influencing the speed at which they can forward packets.

4. In considering the role of routers and hubs in bandwidth consumption, not only the size of these devices' address tables but also such characteristics as their packet filtering and data forwarding rates must be taken into account.

LOOKING AHEAD

In Chap. 6, we move on to investigating the role of protocols themselves in bandwidth consumption.

6

Protocols and Demands on Bandwidth

In Chap. 1, we spent a bit of time dissecting TCP/IP and UDP packets and packet headers, as a means of understanding the effects of those constructs on bandwidth usage. In this chapter, we investigate a corollary idea—the effect of the ways in which protocols operate on that usage.

ARP, RARP, and TCP/IP Packets

When an Ethernet packet is sent, some means must be provided of translating its network routing or IP address to the actual physical address of an end device, that is, to a MAC address. The *address resolution protocol* or ARP handles this job. ARP maps to both IP and hardware addressing, and is therefore key to accurately delivering data.

ARP messages are of either of two types: a request or a reply. An *ARP request* is a message handed to the protocol, which says, in effect, "What's the physical-layer address that corresponds to the IP address 197.201.237.13?" After finding a match in its address resolution table, usually stored in a RAM cache (incidentally an example of the effect of operating systems on bandwidth management), ARP issues its reply, the sending device knows what physical destination it's dealing with, and the actual data goes on its way.

The *reverse address resolution protocol*, or RARP, is ARP's mirror image. RARP, which also sits in the background, receiving requests and issuing replies, translates physical addresses to IP ones.

Obviously this address-translation traffic—only one example of the interpretation that must be done between protocols functioning at different levels of the OSI model—impacts bandwidth. Let's examine the structure of an ARP packet, as a way of illustrating how. An ARP message, whether request or reply, contains fields that identify:

- type of data being transmitted
- hardware type for which the message is intended
- type of protocol address being mapped
- sizes in bytes of hardware addresses and protocol addresses
- whether the message is an ARP request, an ARP reply, a RARP request, or a RARP reply
- sender's hardware address
- sender's IP address
- target hardware address target IP address

As a rule, this ARP/RARP data waits briefly before being processed. A typical ARP cache also records recent mappings from Internet addresses to hardware addresses, usually retaining such entries for about 20 minutes.

Both the UNIX and the Windows NT Server operating systems give you the ability to monitor the ARP cache, and thereby to evaluate the impact of ARP traffic on a network.

Figure 6-1
NT's Performance
Monitor offers dozens
of counter/instance
combinations.

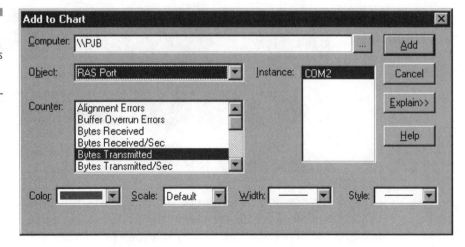

NT and ARP

You might think that Windows NT's Performance Monitor, illustrated
in Fig. 6-1, which tracks and records statistics on everything but the
kitchen sink on a server and network, would be the means through
which you could monitor ARP.

But that's not the case. Under NT, if you want to eyeball ARP, you
must use the command-line utility explained in Fig. 6-2.

In a nutshell, arp lets you look at the current contents of the NT arp
cache. Figure 6-3 gives one such glance.

Figure 6-2
NT expects arp to be
run from the com-
mand line. And don't
let the OS's internal
documentation fool
you. Entering *ARP*
(instead of the cor-
rect *arp*) produces
nothing but a redis-
play of this help infor-
mation.

```
C:\WINNT\system32\cmd.exe                                        _ □

ARP -s inet_addr eth_addr [if_addr]
ARP -d inet_addr [if_addr]
ARP -a [inet_addr] [-N if_addr]

    -a              Displays current ARP entries by interrogating the current
                    protocol data.  If inet_addr is specified, the IP and Physical
                    addresses for only the specified computer are displayed.  If
                    more than one network interface uses ARP, entries for each ARP
                    table are displayed.
    -g              Same as -a.
    inet_addr       Specifies an internet address.
    -N if_addr      Displays the ARP entries for the network interface specified
                    by if_addr.
    -d              Deletes the host specified by inet_addr.
    -s              Adds the host and associates the Internet address inet_addr
                    with the Physical address eth_addr.  The Physical address is
                    given as 6 hexadecimal bytes separated by hyphens. The entry
                    is permanent.
    eth_addr        Specifies a physical address.
    if_addr         If present, this specifies the Internet address of the
                    interface whose address translation table should be modified.
                    If not present, the first applicable interface will be used.

C:\WINNT\system32>
```

Figure 6-3
Only one IP-to-MAC translation is stored in this ARP cache.

Interface: 207.103.234.254 as Interface 3

Internet Addresses	Physical Addresses	Type
207.103.234.255	00-a0-24-24-7a-24	dynamic

UNIX and ARP

An example of output of the UNIX arp command appears in Fig. 6-4.

Figure 6-4 clearly shows the mapping between IP address and Ethernet address for two hosts: mathworks-internet and mathworks-dmz. Figure 6-5 adds to this picture by depicting possible results of using another UNIX network management command, *tcpdump*, to see the details of an arp request/reply pair.

In this example the hosts zippy, dogbert, and gerbil are on the same segment. The host gerbil initiates the tcpdump, watching the ARP transaction between zippy and dogbert.

TCP/IP Routing and the Routing Information Protocol (RIP)

When one TCP/IP host must speak to another, it uses this algorithm to initiate the conversation:

1. First, search for a host address.
2. Next, search for a network address.
3. Finally, search for a route entry.

Figure 6-4
Here's what was going in the arp cache at Worcester Polytechnic a while back.

```
mathworks-internet (144.212.100.1) at 0:0:c:17:e8:85¶
mathworks-dmz (144.212.100.2) at 0:0:c:17:ed:ab¶
```

Figure 6-5
With this combination of arp and tcp-dump, we can even learn about protocol interaction.

```
% arp -a¶

le0 gerbil.mathworks.com 255.255.255.255 08:00:20:18:12:2c¶

le0 dogbert.mathworks.com 255.255.255.255 08:00:20:12:96:bc¶

le0 zippy 255.255.255.255 SP 08:00:20:1b:67:27¶

% tcpdump arp¶

tcpdump: listening on le0¶

09:40:41.835612 arp who-has dogbert (ff:ff:ff:ff:ff:ff) tell zippy 09:40:41.836165 a¶

arp reply dogbert is-at 8:0:20:12:96:bc¶
```

A host address will always be used before a network address in routing TCP/IP packets. That routing relies on a table containing lines like that in Fig. 6-6.

The routing table entry shown in Fig. 6-6 tells us that the route to the Class B network 130.215 is through the interface numbered 130.215.24.56 and through physical interface fza0. So, any packet intended for any host with an address that begins with 130.215 will be transmitted via interface fza0.

This method of conversing across a network is known as *static routing*. It relies on routing table entries made and maintained manually. But there is another, more involved method, called *dynamic routing*, used in more complex networks, which relies, not on administrators, but on a protocol for routing table updates. Such protocols are called, logically enough, *routing protocols*. We investigate them next.

Figure 6-6
This routing table entry offers information on network type, host address, hardware address, and more.

```
130.215 130.215.24.56 U 150 677419902 fza0¶
```

Dynamic Routing and Routing Protocols

Dynamic routing occurs when routers talk to other, usually adjacent routers, informing them of network traffic and route conditions. The vocabulary used in these conversations is made up of routing protocols. We introduced this category of protocols in Chap. 1, in the section "Routing and Bridging." Here, we concentrate on one of the oldest, but still most widely used, routing protocols: the *routing information protocol* or RIP.

RIP is a member of the family of protocols called *interior gateway protocols* (IGPs). As such, RIP allows you to configure many of its operating parameters, among them range of bandwidth used and internetwork delays. Like all routing protocols, RIP does what's called *advertising*. That is, it broadcasts packets to all other routers on the network of which it's aware; these packets are RIP's, and routers' means of keeping one another informed of what's happening on the network. They're also an obvious source of bandwidth overhead. That RIP-related overhead can become even larger, when you consider that whereas RIP advertises every 30 seconds, a router running RIP will wait as long as 180 seconds to receive updates from its peers. If a listening router doesn't get this information in that time, it literally forgets about any routes served by uncommunicative partners, considering those routes to be unavailable for some reason. If the partner remains silent for a total of 240 seconds, the listening router goes a step further, removing from its routing tables all information regarding the quiet device.

Traceroute—Tracing the Path of a TCP/IP Packet

Given the complexity of the interaction of routers, transport protocols, and routing protocols, the need for close scrutiny of that interaction— and for some means of optimizing it in order to optimize bandwidth use—is evident. In this section, we study one means of carrying out that scrutiny.

Traceroute, written by Van Jacobson at Lawrence Berkeley Labs, is a network monitoring and management tool that traces the paths taken by TCP/IP packets as they travel between hosts. Traceroute relies on a protocol called the *Internet control message protocol* or ICMP, as well as the standard Time To Live (TTL) field in an IP header to track packets. Since each router along a packet's path decrements TTL by either 1 or 2, representing, respectively, the number of seconds a given router has held on to the packet, and since the initial value of TTL most commonly is 30, it's possible to use this field to track datagrams with a high degree of accuracy. Another factor improving that accuracy is the fact that any router receiving an IP packet whose TTL is less than or equal to 1 *will not forward* the packet. Instead, the router throws it away, and sends an ICMP *transmission time exceeded* message to the host at which the packet originated. Suitably fastidious, this cross-checking nonetheless is yet another source of bandwidth overhead. And it's at this juncture that traceroute can play an important role.

Traceroute works by using the impatient router's IP address as the source IP address in the ICMP *transmission time exceeded* message. Specifically, traceroute does the following.

- sends an IP message with a TTL value of 1 to the host that is the original packet's intended destination, which causes...

- the first router along the path of this new message to decrement the TTL to 0, to throw away the new message, and to send an ICMP *transmission time exceeded* message to the originally intended destination, thus identifying the first router in the test packet's path; then traceroute...

- sends another IP packet, this time with a TTL value of 2, to identify the next router in the original packet's path...

- and so on, until every router between the original source and destination have been identified

Since hosts don't echo, broadcast, or otherwise advertise ICMP messages, traceroute must use some other means of determining that it has indeed tracked a complete source-to-destination path. So, traceroute sends UDP datagrams to a very high number port that, most likely, won't have anything running through it. This UDP red herring causes the originally intended destination host to generate another type of ICMP message, this one an error message indicating that a port could not be contacted. Traceroute interprets this error message as the signal that it has reached the end of the line.

Figure 6-7
This is only an excerpt, not the entire output of traceroute. You can imagine how many hops the packet being tracked actually took.

```
2·alternet-gw.mathworks.com·(144.212.4.3)·3·ms·3·ms·3·ms¶

3·Boston3.MA.ALTER.NET·(137.39.206.1)·9·ms·7·ms·11·ms¶

4·Fddi0/0.Boston6.MA.Alter.Net·(137.39.99.10)·10·ms·7·ms·14·ms¶

5·Hssi2/0.CR1.BOS1.Alter.Net·(137.39.101.166)·14·ms·8·ms·8·ms¶

6·103.Hssi4/0.CR1.DCA1.Alter.Net·(137.39.30.1)·30·ms·26·ms·26·ms¶

7·Hssi2/0.Vienna6.VA.Alter.Net·(137.39.100.78)·42·ms·29·ms·27·ms¶

8·Fddi0/0.Vienna1.VA.Alter.Net·(137.39.11.1)·32·ms·114·ms·111·ms¶

9·cpe3-fddi-0.washington.mci.net·(192.41.177.180)·112·ms·28·ms·28·ms¶

10·border2-hssi2-0.Washington.mci.net·(204.70.74.117)·34·ms·28·ms·29·ms¶

11·core2-fddi-1.Washington.mci.net·(204.70.74.65)·38·ms·34·ms·31·ms¶

12·core1-hssi-3.Greensboro.mci.net·(204.70.1.129)·43·ms·37·ms·41·ms¶
```

Figure 6-7 gives one example of using traceroute.

We can follow the path of a TCP/IP packet by inspecting each line of any traceroute output. Each such line not only gives the name of every router identified as being on the test packet's path, but also supplies values in milliseconds for three time intervals that represent the delays from the router in question back to the host originating the traceroute request. Such figures can not only help to monitor individual path portions, but even the overall traffic flow across a network. Similarly, the name of each router along a path usually indicates something about what network the packet travels during a given hop, and therefore also gives clues to the physical and geographic path a packet must take.

The Sub-Network Access Protocol

To this point, we've looked at a number of internetwork or intersegment protocols and their effects on bandwidth use and data transfer rates. In this section, we delve a bit deeper, by discussing what must be done to access subnetworks.

During the early 1980s, the IEEE began to define a set of standards for the Physical and Data Link Layers of the OSI model. These standards, part of the IEEE 802 project and initially presented in 1985, contained three Physical Layer standards:

- 802.3 (CSMA/CD)
- 802.4 (Token Bus)
- 802.5 (Token Ring)

as well as one Link Control standard, 802.2. 802.3 or CSMA/CD, the IEEE standard most frequently equated with the term Ethernet, was in turn based on the original Ethernet or DIX standard. This relationship is an important one. While the terms 802.3 and Ethernet are often used interchangeably, they are not truly synonymous. In fact, these standards are incompatible at the Data Link Layer. That lack of interoperability results from a single difference between these standards. In a true Ethernet header, one field is used to store Network Layer protocol type. In an IEEE 802.3 header, this same field contains a value that represents the length of the actual data in a packet.

We won't go into the reasons this incongruity arose. But because of the need for backward and lateral compatibility among a variety of protocols and a steadily increasing number of internetworked hosts, the need for some means of allowing Ethernet and 802.3 to communicate and coexist became apparent. That's when the *sub-network access protocol,* or SNAP, was born.

SNAP's designers realized the desirability of using multiple protocols within a single networking environment, in order to allow nodes more than one means of conversing, and to allow new protocols to be introduced while existing ones still were in operation. These folks realized further that in order to support multiple protocols, some means of distinguishing them had to be provided, so that the correct protocol would be invoked for a given datagram.

SNAP was created by the IEEE to support the coexistence of multiple Network Layer protocols within a single 802 station, as well as to offer backward compatibility with Ethernet. Essentially, what SNAP does is to expand older protocols' 8-bit protocol identification space to a 40-bit protocol ID, and thereby also expand the number of protocols supported. The first 3 bytes or 24 bits of the SNAP Protocol ID are used to store a vendor ID, that is, which organization the Network Layer protocol came from. This vendor ID is actually the same ID used for the first 24 bits of the 48-bit MAC (or Ethernet, or hardware) address associated with a NIC. The last 2 bytes or 16 bits of the SNAP Protocol ID are used to represent the Protocol Type within each vendor, that is, which specific protocol is present. With this combination, the Network Layer protocol type needed by any frame can be completely and correctly determined, and can communicate with Data Link Layer messages. But of course, there's a price. The greater accessibility provided by SNAP carries with it more overhead—those extra 24 bits.

Multimedia and TCP

To this point in our review of TCP/IP protocols and their effects on bandwidth, we've concentrated on traditional, that is, text-based messages and data. But many applications require the capture, storage, manipulation, and delivery of information that must be delivered in a continuous stream, uninterrupted by the kinds of routing and route-management technologies we've spent the bulk of this chapter on. Such continuous transmission, or *streaming,* requires efficient, adaptive protocols if real-time throughput is to be maintained. And therein lies the problem. As we've learned, TCP imposes flow control, error recovery, and more on a data stream. Under TCP/IP, message delivery must be timed. None of these requirements can coexist comfortably with streaming.

A recently developed protocol called the *video datagram protocol* or VDP, on the other hand, attempts to circumvent TCP/IP's real-time shortcomings. VDP offers:

- transmissions that adapt to network load and loss of data due to transmission errors
- similarly adaptive retransmissions
- the ability to determine the nature of and work efficiently with client CPUs

VDP accomplishes this flexibility through:

- UDP datagrams

■ an algorithm for adaptivity, which is based on measurements of transmission latency

■ an algorithm for retransmission-on-demand, which is based on real-time receipt deadlines

■ an algorithm for feedback control, which seeks to ensure a slow increase and fast decrease of strictly administrative messages

Although it is not currently in wide use, VDP offers an example of the direction in which protocols must evolve if the increasingly multimedia and real-time information content of the Internet is to be transmitted efficiently.

IN A NUTSHELL

1. The use of arp and rarp to translate between IP and physical addresses can add to a network's bandwidth requirements.

2. Operating systems like UNIX and Windows NT offer means of monitoring a server's arp cache, and thereby of fine-tuning one aspect of bandwidth performance.

3. Static routing offers less opportunity for incurring bandwidth overhead than does dynamic routing, since the latter relies on additional protocols.

4. RIP, a common dynamic routing protocol, can contribute to bandwidth overhead through its practices of advertising at regular intervals, and of discarding messages and even routes after certain periods of time.

5. Utilities like traceroute can be valuable tools in monitoring performance of a network as a whole, and of individual network segments, routes, and routers.

6. In environments that must ensure backward protocol compatibility as well as future protocol expansion, technologies like SNAP can be employed.

7. In order to aid efficient delivery of streamed information, protocols like VDP must be evolved.

LOOKING AHEAD

In Chap. 7, we take a long and detailed look at the effects of an operating system on efficient use of bandwidth, when we examine UNIX and its networking subsystem, as well as more bandwidth-adept alternatives to that subsystem that have been developed at Rice University.

CHAPTER 7

UNIX and Demands on Bandwidth

In this chapter, we examine the effects on bandwidth consumption of the architecture of a widely used flavor of UNIX, Sun OS. The source for our review is an extensive body of work done by members of the Department of Computer Science at Rice University. That work includes not only a detailed investigation of the impact of this operating system on the efficiency of data communications but also a design for an alternative to the way Sun and more generally BSD UNIX handle that communication.

UNIX

Prior to the Rice studies, work on operating systems' support for and effect on networks focused on delivering a greater percentage of a network's full bandwidth to applications. More recently, resource management for network servers such as LAN servers, firewalls, and HTTP servers has begun to be scrutinized. It's into this latter category that the work at Rice fits. That study points out that although various OSes' policies for process scheduling and memory allocation and swapping seek to ensure efficient operation under various load conditions, the processing of network traffic is frequently little controlled. So, servers that must handle large volumes of network traffic are faced with something of a paradox: how to allocate resources in the orderly way they're accustomed to, to a type of processing which is erratic by definition.

The Rice study goes on to point out that not only UNIX but many other operating systems (among them Windows NT and Windows 95, as it happens) rely on interrupts to handle network processing. Such interrupt-driven management gives highest priority to processing of incoming packets. As a result, not only other network processes but nonnetworked applications can experience decreased throughput and even what the folks at Rice term *resource starvation*. Worst of all, an OS networking subsystem that is strictly interrupt-driven can cause a server to become unstable when its network processing burden becomes too great. In other words, too much network traffic can hang or crash a server.

The BSD UNIX Network Subsystem

Let's shed further light on the operations of a BSD UNIX network subsystem.

BSD and Receiving Data.   Under such a system, an *interrupt* signals the arrival of every new packet. An *interrupt handler,* itself part of another module—the network interface device driver—then does three things:

1. Places the packet in a buffer
2. Queues the buffered packet in the IP queue
3. Issues another interrupt, which causes the IP stack to process the packet

Then, after reassembling fragments into a single coherent IP message if that's been needed, either UDP or TCP is called, as appropriate. Finally, the packet is queued on the socket queue of the socket that has been associated with the packet's destination port.

The sequence just described isn't inviolable, however. Although software interrupts like those that accomplish network processing have higher priority than any user process (thereby ensuring that protocol processing for a given packet will be completed before control of the CPU is returned to any user-initiated process that packet processing may have interrupted), such interrupts have lower priority than hardware interrupts. So, the arrival of new packets can disrupt the processing of earlier ones.

When an application process—let's say, for discussion's sake, a request issued through a client browser—makes a system call requesting the receipt of data to a socket, packet data is copied from the buffer where it was originally and temporarily stored into the application's address space. When this copy has completed successfully, the memory that had been occupied by the temporary buffer is freed up for use by other processes, once again through a system call.

BSD and Sending Data.   When an application tells BSD UNIX it has data to send, that application also writes the data in question to an appropriate socket. Once there, the OS's networking subsystem copies the data into a buffer, which in turn is handed to UDP or IP for transmission. It's also at this point that any disassembling of a message that might be needed takes place.

Whether whole or fragmented, if the intended network interface is busy, IP packets are placed in the interface device driver's queue. All packets, queued or not, get sent on their way by the network interface's interrupt handler.

BSD and Monitoring Network I/O.   Berkeley UNIX has a slightly unusual way of accounting for CPU time consumed by network message processing. Any such processing that results from an application making a system call is considered to be part of that application's use of processor time. Similarly, CPU time spent taken up by software or hardware interrupt handlers is charged, not to the network processing that issued those interrupts and invoked those handlers, but rather to the user process that was interrupted, even if that latter process is unrelated to the network event that caused the interrupt.

Such slightly skewed system accounting can serve to mask bandwidth problems that UNIX can be heir to. So, since BSD doesn't always inform you about those problems, we review them in the next section.

UNIX Bandwidth Problems

BSD UNIX with a standard networking subsystem like that we've just outlined can experience a variety of problems, which the researchers at Rice place in one of four categories. Table 7-1 outlines those categories and the types of networking problems they represent.

BSD UNIX and High Loads

Ordinarily, protocols and distributed applications attempt to prevent a sender process from generating more traffic than the receiver process can handle. Unfortunately, the flow-control mechanisms that seek to accomplish this don't necessarily prevent overload of network servers. Even with efficient flow control in place, a number of factors, such as:

- broadcast and multicast traffic
- incorrect protocol configuration at the client level
- misbehaving applications
- simultaneous requests from a very large number of clients to establish a TCP connection

can cause a server to hang. HTTP, that is, Web servers, can be particularly prone to such problems, since they frequently are subjected to all the circumstances just outlined. Despite TCP's flow-control mechanism, which regulates traffic on established connections, and its limiting retries on connection-establishment requests, TCP effectively imposes no limit on the number of connection requests, allowing those to be bounded only by the capacity of the network.

What's more, distributed applications often provide their own flow- and congestion-control mechanisms. Should these prove inadequate, excessive and even crushing network traffic can result. If such mechanisms happen to coexist and interact with incorrectly configured or inefficiently implemented PC-based TCP stacks, network overload can be further exacerbated.

TABLE 7-1	Type of Problem	Characteristic Circumstances
BSD Networking Subsystem Short-comings	Eager receiver processing	Processing of received packets is strictly interrupt-driven, with highest priority given to the capture and storage of packets in main memory; second highest priority to packets' protocol-related processing; and lowest priority to the applications that consume the messages. Eager receiver processing has significant disadvantages when used in a network server. By assigning highest priority to the processing of incoming network packets, no attention is paid to the condition or the scheduling priority of the receiving application. An arriving packet will always interrupt a presently executing application, even if: ■ the currently executing application is not the receiver of the packet ■ the receiving application is not waiting for the packet ■ the receiving application has a priority lower than or even equal to the currently executing process Bandwidth overhead resulting from dispatching and handling interrupts, and from context-switching, that is, juggling processes running at different priorities, can limit the throughput of a server.
	Lack of effective load shedding	Packet dropping as a means to resolve receiver overload occurs only after significant host CPU resources have already been invested in the dropped packet. Under high load from the network, the system can enter a state known as *receiver livelock*. In this state, the system spends all of its resources processing incoming network packets, only to discard them later because no CPU time is left to service the receiving application programs. In other words, when a server is overloaded, packets may be dropped even after resources have been invested in them. In effect, the server spends much or all of its time processing packets, which it then immediately discards.
	Lack of traffic separation	Incoming traffic destined for one application, that is, for a specific socket, can cause delivery delay and even loss of packets intended for another application. Arriving packets, if they are grouped in bursts, can delay the delivery of an already- or about-to-be-received message by the application for which it's intended. Such delays happen because protocol-related processing of the entire burst or group must complete before any application-initiated process can regain control of the CPU.
	Inappropriate resource accounting	CPU time spent in handling interrupts while receiving packets is charged, not to the program issuing the interrupt but rather to the application that happens to be running when packets arrive. Under UNIX, CPU usage, as monitored and managed by the OS, influences future scheduling priorities. So, some processes may simply, through bad luck, be bumped now and in the future.

Even tools designed to protect or aid other aspects of a network can contribute to inefficient bandwidth use. For instance, a packet-filtering application like a firewall establishes a new TCP connection for every flow that passes through it. Since these connections are under the supervision of the server operating system and must function at the direction of its networking subsystem, they are subject to all the interrupt- and flow control–related problems we've cited to this point in this chapter. So, an excessive flow establishment rate might seize an unfair share of server resources, or interfere with other communications processes. Similar problems can occur in systems that by their nature simultaneously execute several server processes. For example, Web servers can associate one server process with each connection made to them. And, of course, applications like those involving multimedia, which have stringent scheduling and coordination requirements, can little afford to be subjected to scheduling anomalies that can result from an overburdened network server.

At this point, two things should be clear:

1. that a network server must, if it is to function efficiently, be able to control its resources so as to remain stable under high loads

2. that traditional interrupt-driven network subsystems aren't very good at accomplishing this

LRP

As an alternative to the various shortcomings the standard BSD network subsystem can demonstrate, the researchers at Rice have devised a new subsystem architecture based on what they call *lazy receiver processing* or LRP. Their proposed subsystem overcomes BSD's networking problems through the combined use of several techniques.

- The IP queue is replaced with a per-socket queue that is shared with the network interface.

- The network interface demultiplexes incoming packets according to their destination socket and places the packet directly on the appropriate receive queue. Packets destined for a socket with a full receiver queue are silently discarded (early packet discard).

- Receiver protocol processing is performed at the priority of the receiving process.

■ Whenever a protocol's semantics allow it, protocol processing is performed lazily, that is, by means of an end-user-initiated process such as a browser's carrying out a system call.

This design attempts to ensure that packet receipt and processing will not interrupt any process that might be running when the packet arrives, unless the process that acts as receiver has a priority higher than that of the executing process. Also, under the Rice design, a network interface sorts incoming traffic by destination socket, allowing it to place packets directly into per-socket receive queues. Such sorting not only speeds up overall processing of network traffic but can also, when taken together with LRP's application-driven priorities, inform the network interface about application processes' ability to keep up with traffic arriving at a socket. Such information can be valuable to an interface, since it permits that component to ignore packets destined for a given socket until some of the packets already queued to that socket have been dealt with by the applications for which they were intended. In other words, an LRP interface can effectively work through its load without gobbling up too high a portion of server resources. As a result, a server running LRP is more likely to stay alive during high-load periods, and can even demonstrate decent throughput at such times.

Finally, a network interface operating under the direction of LRP, because it separates received traffic helps eliminate interference among packets intended for separate sockets, while LRP's doing away with a shared packet queue minimizes the chance of a packet's being delayed or dropped.

Early Demultiplexing. LRP isn't the first network subsystem design to rely on early demultiplexing. That technique has been used in many blueprints, for purposes as varied as:

■ supporting application-specific protocols

■ avoiding duplicating data

■ keeping the "real" in network real-time communications

Other of LRP's techniques have been used before as well. For instance, demultiplexing at the network adapter has been employed as a means of achieving high-bandwidth user-level communication. But prior to LRP, these and similar schemes hadn't been used in conjunction with delaying the processing of incoming packets. The two methods most important to, and with the greatest impact on, LRP—lazy protocol processing at the priority of the receiver, and early demultiplexing—must both be present in order to allow a networking subsystem to attain optimum performance, especially when dealing with heavy loads. Early demultiplexing alone cannot ensure such performance.

The researchers at Rice offer this example. Consider a system that combines traditional eager protocol processing with early demultiplexing. Even with the presence of early demultiplexing, packets are often dropped immediately because their destination socket's receive queue is full. Further, and as a result, such systems still can demonstrate poor performance under heavy loads, such as when a rash of control messages or corrupted data packets induces server lock, because processing of these packets still isn't efficient in placing them in the proper socket's queue.

Sockets and Channels.   If sockets can be considered virtual connections, a network interface channel can be thought of as something similar: a virtual communications path. A *network interface channel* is a data structure that is shared by the NIC and the operating system kernel. Such a channel consists of:

- a receiver queue
- a free buffer queue
- state variables

Anytime you bind a socket to a port, a network interface channel is therefore by definition created. All traffic, whether unicast or multicast, which either originates at or is intended to arrive at a given socket, must travel that socket's *network interface* (NI) channel.

LRP requires that all NICs be able to identify the destination socket of an incoming network packet, so that the packet can be placed on the correct NI channel. Although many commercial network adapters, especially the higher-speed models, have an onboard CPU, and are therefore capable of demultiplexing, not all are as skilled. So, LRP assumes nothing. It determines if the adapter of the moment can demultiplex. If so, it allows that process to take place at the NIC. If not, as, for example, when LRP detects any one of a number of models of inexpensive Fast Ethernet adapters, demultiplexing is instead done by the network driver's interrupt handler, and is referred to as a *soft demux*. With a soft demux, some host interrupt processing still must occur when demultiplexing incoming packets. Such overhead, though, does not significantly affect server performance under heavy loads. And soft demuxing has a more explicit advantage. It is hardware-independent.

LRP's soft demultiplexing makes as few demands as possible on server resources. For instance, it does no dynamic memory allocation and uses no timers. This design for leanness makes LRP easier to integrate into a NIC's host interrupt handler. What's more, LRP's demuxing routine handles all types of TCP/IP packets efficiently, even IP fragments.

LRP and UDP. The *user datagram protocol* or UDP is what's known as unreliable. This means that UDP neither delivers packets in the order in which they were sent, nor retransmits damaged or dropped packets. So, in dealing with UDP, LRP faces a different scenario. On the transmit side, there's little difference between IP and UDP processing, whatever the nature of the networking subsystem. But on the receiving side, here's what happens.

The NIC determines the destination socket of the incoming packet. Then, it shifts that packet to the appropriate NI channel queue. If that queue is full, the packet is discarded, without any complaints, this being UDP. So, when an application process makes a receive system call on a UDP socket, the system may or may not find anything there. If it does, the packet is removed; the IP input function is called, and it in turn calls the UDP input function. Or, as the folks at Rice put it, the processed packet is eventually copied into the application's buffer.

We need to emphasize several things about LRP's UDP receiver processing. First, packet processing doesn't even begin unless:

- an application is waiting for the packet
- the packet has arrived
- the application is scheduled for the CPU, rather than waiting

Second, when the rate of incoming packets exceeds the rate at which the receiving application can consume the packets, the channel receive queue fills, and the NIC drops packets, *before* significant host resources have been consumed. As a result, throughput can reach and remain at maximum even if the arrival rate for packets continues to shoot up.

Although LRP thus makes delivery of UDP packets as efficient as it can, it cannot solve or ameliorate another problem associated with UDP transmissions—*latency,* that is, delivery delay. Remember, UDP is an unreliable protocol. For such protocols, the only way to reduce latency is to increase the amount of CPU time allocated to processing arriving packets. If you've got a multiprocessor machine, you've got a leg up on the situation. If not, you must take the tack suggested by the Rice researchers, and designed into LRP by them. A kernel thread must check NI channels regularly, and immediately cause an otherwise idle CPU to process any queued UDP packets detected.

LRP and TCP.   LRP doesn't have its hands quite as full when working with reliable, flow-controlled protocols like TCP. In dealing with TCP packets, LRP's approach closely resembles that of UNIX's default network subsystem. Data that an application needs to transmit is queued to the appropriate socket. Some of it is transmitted immediately, whereas the balance may have to await the sending machine's receiving acknowledgments or timeouts.

Because TCP is flow controlled, a receiving machine, not a sender, controls the pace of transmissions, by means of acknowledgments. In such a scenario, high throughput therefore requires efficient handling of these acknowledgments. To effect such handling, LRP carries out receiver processing for TCP sockets asynchronously, if that's what's needed to move both acknowledgments and data. In this way, packets arriving on TCP connections can be processed more smoothly, without the need for either receive system calls or too long a wait for an acknowledgment. Further, the kernel thread assigned to application processes that use TCP sockets benefits from the fact that protocol processing always runs to completion, and that therefore no separate runtime stack need be set up for the asynchronous processing thread. That thread relies instead on a single stack, thus once again reducing overhead.

LRP and Other Protocols.   Some network packet processing can't be directly associated with any application. TCP/IP family members such as the *address resolution protocol* (ARP) and the *reverse address resolution protocol* (RARP) fall into this category. LRP charges CPU time involved in processing such packets to daemon processes, which are proxies for protocols. These daemons are assigned an NI channel. Packets of a particular protocol are associated with a particular daemon, and can therefore be demultiplexed directly onto the appropriate channel.

Experiments with LRP

Rice's computer science department didn't just theorize about improving the UNIX networking subsystem; they thoroughly tested their design. Their testbed included a number of Sun Microsystems SPARCstation 20 model 61 workstations, equipped with a 60-MHz SuperSPARC + processor, 32 MB of RAM, SunOS 4.1.3_U1, and a 155 Mbps ATM LAN connection by means of a FORE Systems SBA-200 network adapter. The SBA-200 carries an Intel i960 processor that can handle such tasks as reassembly of protocol data units. But it's important to note that LRP isn't wedded either to any one adapter or to ATM networks. LRP in its soft demuxing incarnation can coexist with any network and any NIC.

Given this preamble, LRP was implemented by modifying the 4.4 BSD-Lite distribution's default TCP/UDP/IP network subsystem. The enhanced subsystem was included into the SunOS kernel as a loadable kernel module and attached to the socket layer. (A special device driver was also developed to talk to the FORE network adapter.) The researchers at Rice chose the 4.4 BSD-Lite networking subsystem because of its performance and the availability of its source code. To evaluate packet demultiplexing in the network adapter, they used firmware developed for the SBA-200 by Cornell University's U-Net project. This firmware performs demultiplexing based on the identification of ATM virtual circuits. Given the complexity of their overall undertaking, and of the task of modifying such firmware, the researchers at Rice deserve an award for understatement when they state simply that the resulting implementation of NI-LRP is fully functional.

Finally, the machines on which LRP was tested ran in multiuser mode, but were not actually handling multiple simultaneous user sessions.

Throughput and Latency.   Latency was measured by bouncing a 1-byte message back and forth between two workstations 10,000 times. The total elapsed time was then divided by that figure of 10,000 to come up with an average round-trip latency. The researchers measured UDP throughput primarily by disabling UDP checksumming, and TCP throughput by transferring 24 MB of data over a socket whose send and receive buffers had each been set to 32 KB.

Under these test conditions, LRP demonstrated improved performance under heavy loads as compared to that offered by the unmodified BSD networking subsystem, while at the same time keeping pace with that subsystem in low-load performance.

Performance Under Heavy Loads.   In this test at Rice, a client process sent 14-byte UDP packets at a predefined, fixed rate to a server process on a remote machine. That server process discarded the packets immediately upon receiving them.

Under such conditions and the default 4.4 BSD network subsystem, throughput increases with load, but only up to a maximum of 7400 packets per second. Beyond that, throughput decreases, even to the point of the server's risking hanging, at about 20,000 packets per second. Under NI-LRP, on the other hand, throughput increases up to a maximum of 11,000 packets per second. What's even better, LRP maintains that level of throughput even when offered load increases further. These results confirm that LRP effectively sheds load to the network interface, before undue host resources have been gobbled up by the need to deal with excess traffic.

Under LRP with soft demultiplexing, that is, demultiplexing carried out by a server's interrupt handler, throughput levels off at about 9760 packets per second, and falls a bit below this rate with increasing offerings. This backsliding can be attributed to the overhead inherent in host software-based demuxing. So, while NI-LRP eliminates server lock, software-based or SOFT-LRP might only delay it. But there's a silver lining. Although they tried, the tests at Rice could not generate packet rates high enough to lock up the SOFT-LRP kernel.

That kernel was also compared for performance rates to a kernel that had been modified only to include early demultiplexing, and not lazy receiver processing. The latter kernel, like SOFT-LRP, demultiplexes in the interrupt handler. Because of such early demultiplexing, some UDP lookup is eliminated in both kernels. The early-demux kernel, once again like LRP, outperformed the default BSD network subsystem under heavy loads, by employing early packet discard. Also, there's little to distinguish the LRP from the early-demux kernel in rate of decline under overload. But here's the kicker. Overall throughput of the early-demux kernel is between 40 to 65 percent of SOFT-LRP's throughput in areas of the network experiencing traffic jams.

Overall Throughput.   Both variants of LRP offer better throughput than either the conventional 4.4 BSD networking subsystem or an early-demux kernel. NI-LRP has a maximum delivery rate 51 percent higher than that of standard BSD, moving packets at 11,163 per second compared to BSD's 7380 per second. SOFT-LRP, although not quite as speedy as its NI sibling, nonetheless is about 32 percent faster than BSD; SOFT-LRP moves about 9760 packets per second.

Investigating LRP Throughput. LRP kernels were examined with instrumentation in order to add detail to the picture of the subsystem's throughput gains. As a result, it was found that SOFT-LRP's *maximum loss-free receive rate* or MLFRR exceeded that of 4.4 BSD by 44 percent, or in raw numbers by 9210 as opposed to 6380 packets per second. Since both 4.4 BSD and LRP drop packets at the socket queue or NI channel queue when offered packets exceed this MLFRR, we can conclude that early packet discard doesn't play a role in MLFRR performance differences. Similarly, since all of BSD, early-demux, and SOFT- and NI-LRP kernels rely on the same 4.4 BSD networking code, and since the device driver and demultiplexing code used in the early-demux and SOFT-LRP kernels are identical, we can't consider any of these as contributing significantly to LRP's throughput gains. Rather, those performance gains must be largely due to factors such as better handling of interrupts and improved access to memory.

Latency Under High Loads. To test the latency that a client experiences when trying to reach a server process running on a machine, which in turn is experiencing high network load, the folks at Rice came up with this scenario.

A client process on machine A exchanges a short UDP message with a server process on machine B. At the same time, machine C zaps machine B with a high level of UDP packets; B discards these packets as they arrive. For all three experimental kernels, measured latency, as might be expected, varied with background traffic rate. The need to process background traffic delays both request processing and response transmission between A and B, thus causing round-trip delay to increase. How much will a packet's round-trip time go up under such conditions? Instrumentation-based observations confirmed that the extent of the increase in round-trip delay depends on both the rate at which background packets show up, and the length of the interruption caused by each such arrival. Because of its larger overhead, made up of both hardware interrupts, software interrupts, and protocol processing, 4.4 BSD shows the highest rate of increase in round-trip delay. SOFT-LRP's reduced interrupt overhead provides it a lesser increase, whereas NI-LRP's reliance on only hardware interrupts in combination with minimal packet processing allows it to experience very little round-trip delay increase.

The Role of the UNIX Process Scheduler.   Any UNIX scheduler assigns processing priorities based on a process's recent CPU usage. As a result, processes that have been waiting for a packet to arrive, that is, which haven't yet consumed significant CPU time, tend to be more important, in the scheduler's scheme of things, than processes that have been interrupted by packets arriving. This practice of the scheduler bears directly on the various kernels' performance in the heavy-load tests conducted at Rice. At low packet-arrival rates, machine B, or what's called the *blast receiver,* will always exchange its UDP message with machine A rather than turn to handling the stream being zapped at it by machine C. However, if the arrival of a blast from C at B happens to interrupt directly the receipt of a packet from A, the scheduler almost always tells the CPU to deal with the background packet, thus delaying handling the A/B message significantly. But when overall arrival rates of about 6000 packets per second are reached, the client process that receives these background packets approaches its processing limits, and as a result, its priority decreases, and the scheduler can choose to return the CPU to the interrupted A/B conversation. In this way, the intrusiveness of background traffic can be lessened or even eliminated at high overall traffic rates.

Preventing Latency from Increasing.   Rice also measured LRP's ability to keep latency at an acceptable level during bursts of network traffic. In this test, a client periodically transmitted a sequence of packets in immediate succession to a server process on a second machine. The server process responded with an acknowledgment for the first packet. When this acknowledgment arrived at the client, and after a small delay in order to ensure the server had enough time to grab all the packets in the previous burst, the client transmitted another burst.

The interval between the start of a burst transmission and the receipt of an acknowledgment by the client was measured and averaged for 1000 transmissions. Under the 4.4 BSD networking subsystem, latency proved to increase linearly with the size of the burst. On the other hand, under both LRP kernels, packets, which are processed in the order they arrive, with earlier packets being delivered to an application before protocol processing for any subsequent packets takes place, experienced little increase in latency. That demonstrated by NI-LRP was attributed to the overhead of handling hardware interrupts for subsequent packets in a burst train. SOFT-LRP's slight increase in latency during burst was felt to be due to the overhead inherent in demultiplexing packets in a burst.

Modeling the Real World. LRP's designers have also attempted to simulate the effects of the enhanced networking subsystem on a mix of workloads that might typically be found on network servers. They loaded three processes onto a server machine. The first, which they termed the *worker,* carried out memory-bound computation in response to a *remote procedure call* (RPC) from a client. This calculation ate up about 11.5 seconds of CPU time, and used about 35 percent of the server's second-level cache. The other two server processes were less demanding of system resources, since they did only short computations in response to RPCs. With all this going on, a client on a separate machine sent an RPC to the worker, and, while this request was still unfulfilled, to the other two server processes. All this was done in such a way that each server had several outstanding RPCs at any given moment, in order to try to ensure that the server processes generated by these RPCs never ignore network-based requests. Further, this queue, if you will, of outstanding processes was distributed as much as possible uniformly over time, so as to try to ensure that no correlation would arise between server scheduling of the RPCs, and the times at which those requests were forwarded by the client. Given these parameters, it was found that a server's throughput, in this case consisting of RPCs completed and worker completion time, was lowest with traditional BSD, higher with SOFT-LRP, and highest with NI-LRP, thereby confirming that LRP's designers had accomplished one of their major goals: to increase throughput under high load.

Traffic Separation. More real-world tests were done of LRP. To further demonstrate traffic separation, the subsystem's designers set up a series of tests in which a machine running a SOFT-LRP kernel was configured to act as a Web server, using NCSA httpd revision 1.5.1. This souped-up server proved to be far more stable under heavy loads than one based on unmodified BSD. As part of the test, several browser clients running from a single machine contacted the HTTP server, while at the same time a program running on a third machine sent packets to a separate port on the server machine at the rate of 10,000 packets per second. The HTTP server that relied on 4.4 BSD hung up under these conditions, not responding to any HTTP requests, and not even accepting input at the console. Under SOFT-LRP, however, the console remained available, and the HTTP server continued to respond to browser requests.

In another, similar test, eight HTTP clients on a single machine continually asked the server for the same 1300-byte document, while a second client, on a different machine, sent fake TCP connection establishment requests to a dummy TCP server running on the same machine that housed the HTTP server. (These requests were dummies in that no session was ever established as a result of them.) Also, in order to allow BSD's performance in this test to be as free as possible of inherent limitations, especially that related to its practice of using lookup functions in dealing with HTTP servers, its TCP TIME_WAIT parameter was set to 500 milliseconds, rather than the BSD default of 30 seconds. Finally, these tests were run for long periods of time, and even involved LRP's doing a lookup, to eliminate any experimental bias that might arise from the greater efficiency of LRP's early demultiplexing.

Under these conditions, the number of HTTP transfers completed by all clients when talking to the BSD HTTP server dropped sharply as the rate of background requests increased. This result is understandable when you realize that BSD's processing of TCP connection requests starves an HTTP server as far as CPU time is concerned. What's more, at overall arrival rates for connection requests in excess of 6400 packets per second, such packets are dropped by BSD's shared IP queue, resulting in the loss of both TCP connection requests from HTTP clients and traffic on established TCP connections.

LRP fared much better under these conditions. At 20,000 background requests per second, the LRP-based HTTP server still functioned at about 50 percent of the maximum throughput of which a hit was capable. Under LRP, established TCP traffic, HTTP connection requests, and dummy TCP requests all were demuxed onto separate NI channels, and therefore didn't interfere with one another.

Heavy Levels of HTTP Requests.   Finally, a set of experiments that generated HTTP requests in excess of the server's capacity from a small number of client machines was conducted at Rice. Again, to avoid bias due to the known inefficiency of protocol lookup in BSD, a version of the BSD kernel was implemented that used hashed lookup.

The makeup of the request rate of the 4 client machines was varied by altering the total number of outstanding requests per client, and the maximum duration for which a client waited on a connection establishment request before trying again. Under these conditions, the throughput of an HTTP server based on vanilla 4.4 BSD dropped sharply as the rate of HTTP requests increased. Because of unmodified BSD's linear-search lookup algorithm, and the large protocol lookup table that usually is found on Web servers, lookup-based connection request processing is voracious in terms of CPU time, once again causing an HTTP server to go hungry where CPU resources are concerned. The situation improves somewhat when 4.4 BSD uses hashed lookup. Under this scheme, overall server throughput decreases less sharply as HTTP requests rise. Finally, under LRP, throughput decreases only slowly, because connection establishment requests don't starve out the processing of packets for established connections.

Related Work

Environments, like that involving the state of California's 1994 Election HTTP server, also demonstrate many problems, like those documented by LRP's designers, encountered by conventional network subsystems when these must act as HTTP servers under heavy load. Some researchers have gone so far as to suggest that customization like that employed by the staff at Rice may be the only way to ensure adequate performance by busy servers. Others have come up with differing techniques for improving the overload behavior of an interrupt-driven network architecture. The overload stability of servers operating under such paradigms appears to be comparable to that of LRP, at least in most areas.

One thing many researchers in this field agree upon is the importance of early demultiplexing to high-performance networking. Early demuxing, that is, demultiplexing that takes place immediately at the network interface, is seen by many as critical, since it allows:

- user implementations of network subsystems
- more effective placement of data in memory
- realistic resource accounting in the network subsystem
- stability of network subsystems under heavy loads

IN A NUTSHELL

1. The networking subsystems of many widely used operating systems, including popular flavors of UNIX, are, like the OS as a whole, interrupt-driven.

2. The standard networking subsystem of BSD UNIX signals the arrival of every packet with an interrupt.

3. When transmitting, the standard networking subsystem of BSD UNIX queues packets to a network interface, whose interrupt handler in turn controls actual transmission of those packets.

4. BSD's standard networking subsystem charges CPU time required by network packet processing, not to the interrupts or handlers that requested that processing, but rather to the user applications that generated the network request, thereby possibly unfairly affecting subsequent allocation of the CPU to these applications.

5. When running under UNIX, HTTP servers are particularly prone to hanging because of circumstances like:

   - broadcast and multicast traffic

   - incorrect protocol configuration at the client level

   - misbehaving applications

   - simultaneous requests from a very large number of clients to establish a TCP connection

6. Many researchers feel that early demultiplexing, like that employed in Rice University's lazy receiver processing (LRP) redesign of the BSD networking subsystem, is critical to efficient performance for high-traffic networks.

7. Some researchers have suggested that customization like that done to produce LRP at Rice may be the only way to ensure adequate performance by busy servers.

LOOKING AHEAD

Having dissected the role of UNIX in bandwidth consumption, we turn in Chap. 8 to putting that of Windows NT and Windows 95 under the microscope.

CHAPTER 8

Windows NT, Windows 95, and Demands on Bandwidth

In this chapter, we examine the effects on bandwidth consumption of:

- Windows NT 4.0
- Windows 95

Windows NT Server 4.0

NT Server 4.0, Microsoft's contender for the title of multiuser, multitasking champion, presents a number of constraints upon bandwidth. Like UNIX, NT has a networking subsystem that is interrupt-driven. In addition, some of the bandwidth overhead NT experiences relates to the protocols this OS makes available. So in this section we review those protocols in some detail.

NT's Protocols

Windows NT supports TCP/IP, NWLink, and NBF as transport-layer protocols, and the DLC protocol, which does not provide transport layer services. By *NWLink*, Microsoft means its drivers that provide communication with Novell's IPX/SPX suite. By *NBF*, the company means drivers to accomplish interface with the IBM-designed NetBEUI protocol family.

Under Windows NT, NWLink and TCP/IP load by default, so that their initial setup cannot be customized. However, once the initial installation is complete, administrators can configure these protocols more closely. Microsoft recommends that you do so, especially if your environment offers more than one protocol suite. The reasoning behind this recommendation is sound; as one Microsoft white paper put it, running multiple protocols usually results in both higher memory requirements and more complex configuration and administration, either of which, in the context of bandwidth, can make for less than efficient use.

NOTE: *The same can be said of any operating environment. So, what we discover here in reviewing the relationship between Windows NT and bandwidth also applies to other operating systems.*

Windows NT Transport Driver Architecture

Windows NT Server uses as a paradigm the *Network Basic Input/Output System* or NetBIOS standard, originally developed for IBM in 1983. NetBIOS specifies two unique objects.

The Session Layer API. Any NetBIOS-compliant protocols make use of a Session Layer application program interface that defines how end-user applications must submit requests for network services to protocols. Such requests, handled as NetBIOS commands, are forwarded to protocols by what are known as *network control blocks* or NCBs. Another protocol, dealing with session management and data transport, and called the *NetBIOS frames protocol* (NBFP) exists at both the Session and Transport Layers to establish network I/O for the NetBIOS command set.

Here's where it begins to get tricky. An application program that uses the NetBIOS interface API to talk to a network can work with any protocol driver that in turn uses the NetBIOS interface as we've just defined it. The trouble is, neither TCP/IP nor IPX/SPX does so out of the box. For them to run in conjunction with NetBIOS, some means of mapping every NetBIOS interface command to an area of these suites must be provided.

Windows NT transport drivers don't do so. Rather, they employ what Microsoft considers a more flexible tool, the *transport driver interface* or TDI. Windows NT uses a NetBIOS Emulator to map NetBIOS commands to TDI commands and events. In other words, an original Net-BIOS sequence must be translated into TDI, since this is the only interface used internally by Windows NT's network subsystems to communicate with transport protocols.

TDI-based processes complicate the NT protocol picture further, since transports that do not include NBFP, which as we've seen means most of the protocol world, must use a NetBIOS compatibility layer to translate addresses formatted for NetBIOS to their own address format, and to transfer messages over the transport's native protocols.

More Detail on NT's Transport Driver Implementation

Except for its Data Link Layer drivers, all of NT Server 4's transport control utilities use a dual interface, working with the TDI interface to communicate with calls from applications and from the NT Server itself, while also using a *network driver interface service* or NDIS to talk to drivers for NICs. Not only does this dual-driver scheme itself introduce overhead; the nature of the transport protocol paired with NDIS can increase that draw on bandwidth. In this section, we examine this duality further.

NBF (NetBEUI) Under Windows NT.   Windows NT's NetBEUI Frame or NBF transport driver complies with IBM's NetBEUI 3.0 specification, and, as important to our discussion, fully implements NBFP, therefore needing no NetBIOS compatibility layer. That being the case, and because NetBEUI packets carry no headers, communications between NetBEUI stations, such as those running under older versions of Windows 3.x and Windows 95, and Windows NT Server incur less transmission delays than those between NT and, for example, TCP/IP clients.

TCP/IP.   Windows NT offers a TCP/IP implementation that includes the most significant protocols in this family, among them TCP and IP themselves, as well as UDP, ICMP, and ARP. Although this reliance on the TCP suite gives NT a high degree of interoperability, it also contributes to bandwidth overhead, since the Microsoft design for the TCP/IP transport driver in Windows NT 3.5 and higher uses a NetBIOS compatibility layer over TCP/IP itself.

NWLink (IPX).   As it does with TCP/IP, Microsoft implements many members of another protocol family, Novell's IPX/SPX. Among the NetWare protocols Microsoft makes available are IPX, SPX, RIPX, and NBIPX. But as it did in implementing TCP, NT Server uses NetBIOS, this time in conjunction with IPX, thereby increasing demands on transmission resources by virtue of the need for these components to converse.

Comparing NT Transport Drivers

In this section, we compare Windows NT's transport protocols according to several criteria:

- configuration and administration
- handling of network segmentation
- handling of network status reporting
- management of network traffic
- performance
- routing

> **NOTE:** *Keep in mind that, however well any of the three Windows NT protocols, or any protocol, performs in any of these areas in theory, their performance in your environment will be affected by such additional factors as:*
>
> - *the size of your network*
> - *whether your network consists of one or a number of geographical locations*
> - *whether your network is largely homogeneous or predominantly heterogeneous*
> - *whether, and how heavily, your network interacts with the Internet*

Configuration and Administration. As Microsoft terms it, all three protocol suites offered under Windows NT Server are *self-tuning*. This is so much the case that both Microsoft's own, and other, documentation on configuring these protocols often urges administrators to avoid changing parameters, particularly those Microsoft classifies as advanced, from their self-tuned defaults. However, as Fig. 8-1 illustrates, many of these parameters are among those that, as we've discovered in earlier chapters, can have a significant effect on bandwidth consumption.

Whereas NT Server does allow the administrator to manually configure certain protocol parameters, it simultaneously discourages him or her from doing so, by emphasizing:

- the complexity of the TCP/IP naming scheme
- the need to identify a default gateway or router for every station on a network
- the role of the *dynamic host configuration protocol* or DHCP as an alternative to these TCP complexities

Even Microsoft admits, however, that using DHCP as an alternative to grappling with the details of the TCP/IP suite has its drawbacks. Although this protocol and its dynamic negotiation of client addressing does allow the administrator to forgo manual setup of clients, DHCP introduces bandwidth overhead not only simply through its presence, but also because it requires the definition and maintenance of DHCP servers.

NWLink, NT's IPX/SPX implementation, has an advantage over the OS's TCP. Because NWLink uses the MAC address of a client NIC as the

Figure 8-1
Figure 8-1
These advanced
properties of
TCP/IP clearly can
affect transmission
efficiency.

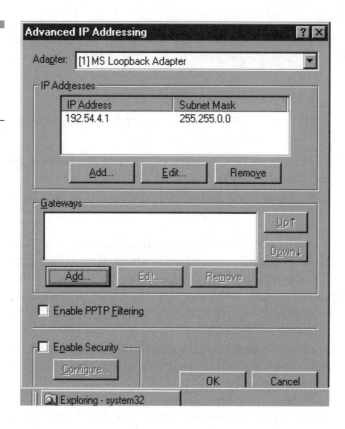

node component of an IPX address, clients almost never need be configured manually. However, as you can see from Fig. 8-2, configuring NT Server to accept transmissions from these clients isn't as simple.

Handling Network Segmentation. Segmenting a network into small sections not only makes both sections and network-at-large more manageable, but can also help to reduce overall network traffic. However, this subdivision, because it requires a more complex addressing scheme, also may introduce its own demands on bandwidth.

NetBEUI node addresses, as we found in the section "NBF Under Windows NT" earlier in this chapter, fully implements NBFP, and therefore offer no defining hierarchy within an address, such as may be used under TCP/IP to distinguish between, for example, individual nodes on a segment within an interconnected network. TCP itself, of course, through its famous four-part addresses, allows the precise location of nodes within even very complex, highly subdivided nets. IPX lies some-

Figure 8-2
Take a look at what
must be done to set
up NWLink at the
server end.

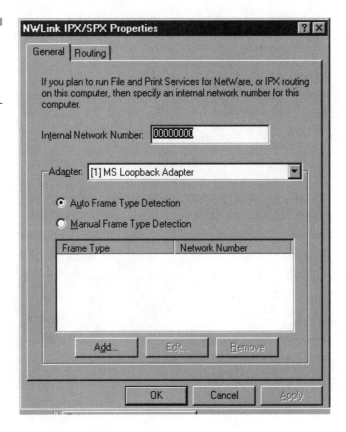

where between these two, with its two-part naming scheme. That scheme, however, employs no hierarchy, and therefore doesn't allow the establishing of subnets.

Network address resolution has another impact on traffic, particularly in multiprotocol environments like those NT Server can offer. Because both broadcast and multicast traffic by definition affect not only local segments but an entire network, such transmissions, when also having to deal with a combination of address resolution schemes such as that presented by Windows NT, can be particularly demanding of bandwidth.

Where does all this leave Windows NT Server? Because it:

- must deal with three protocol suites whose addressing and architecture differ significantly from one another

- may have to handle conversations between stations or entire segments whose addressing and even routability are incompatible

**Figure 8-2
(Continued)**
Take a look at what
must be done to set
up NWLink at the
server end.

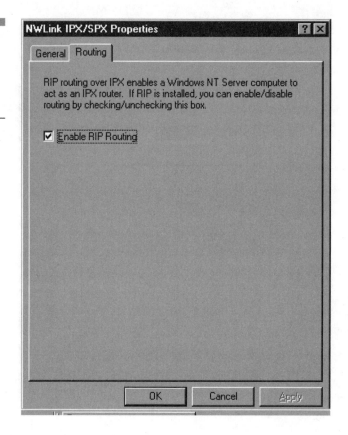

- leans heavily on DHCP to avoid some of the drudgery and complexity of TCP/IP configuration, thereby adding another level of complexity to intersegment exchanges

the OS, through its attempts to offer so wide a range of communications capabilities, also presents the potential for significant bandwidth overhead.

HANDLING NETWORK STATUS REPORTING AND MANAGING TRAFFIC. Regular and frequent monitoring of even a simple network is critical to fine-tuning its performance. When a segmented or otherwise multiprotocol environment must be managed, the ability to eyeball status from the broadest, that is, the overall network level, to the most narrow, or node, levels, is even more so.

Figure 8-3

Performance Monitor can set up a slew of TCP/IP or IPX/SPX conditions to track.

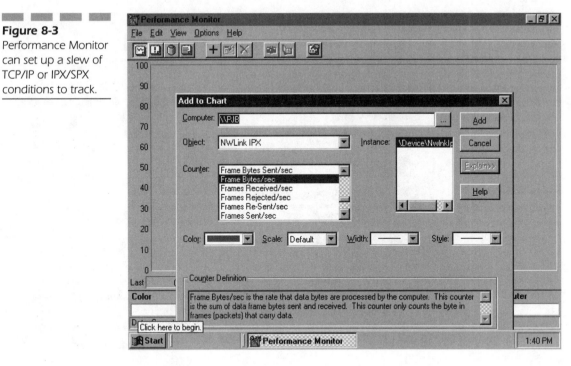

In and of itself, Windows NT Server offers two network monitoring tools: (1) *Event Viewer,* which automatically reports both standard and atypical server events and (2) *Performance Monitor,* illustrated in Figs. 8-3 and 8-4, offers detail that Event Viewer does not.

Among the IP parameters Performance Monitor can eyeball are:

- datagrams received but discarded
- datagrams received that contained header errors
- datagrams received per second
- datagrams transmitted per second

and over two dozen more. IPX factors the Monitor can track, some of which appeared in Fig. 8-3, include:

- total bytes sent per second
- connections cancelled
- connections open

Figure 8-4

Performance Monitor, as this illustration of its Remote Access Service tracking capabilities shows, not only lists but can also briefly explain the counters it uses.

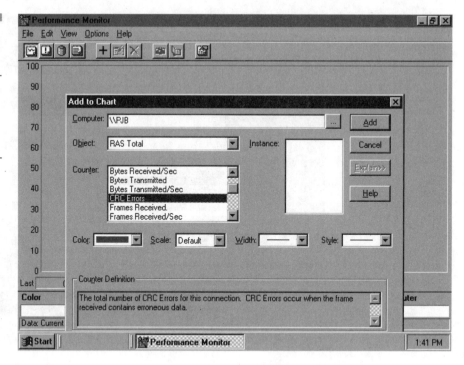

- failures in a link
- piggyback transmission acknowledgments

NOTE: *The build of Windows NT Server 4.0 that we used in preparing this book made exactly the same Performance Monitor counters available for overseeing NetBEUI traffic as it did for tracking IPX.*

However, NT Server relies on ICMP as it is implemented in particular routers for notification that problems such as a node's being unreachable have occurred. This fact, as well as its multiprotocol capabilities, can cause the OS to provide less than thorough network usage statistics under some circumstances. For example, neither NBF nor IPX gathers any network status information. What's more, IPX has no management protocol analogous to TCP's ICMP. As a result, IPX routers can't tell a sending station things like:

- a destination is unreachable

- traffic is heavy on a segment (IPX or other)
- transmissions must be timed differently

ROUTER BROADCASTS.   As we learned in Chap. 5, routers themselves contribute to demands on bandwidth, largely through their practice of broadcasting to one another. Such exchanges don't relate to NetBEUI. Since it isn't routable to begin with, this protocol has no effect on either router broadcasts or those broadcasts' consumption of bandwidth. Both IP and IPX routers, however, keep their routing tables up-to-date RIP broadcasts to every port they're aware of. IP broadcasts every 30 seconds. IPX is a little less talkative; it broadcasts every 60 seconds. What's more, any NetWare file server must also carry out routing, and so also broadcasts RIP messages. But perhaps the biggest contributor to bandwidth overhead generated by router broadcasts in multiprotocol environments like that offered by NT is the fact that IP RIP can't talk to IPX RIP. As a result, such environments often transmit redundant RIP broadcasts.

SAP BROADCASTS.   In pure NetWare networks or segments, IPX servers notify nodes of server availability and resources with another type of broadcast, the *service advertising protocol* or SAP. These broadcasts transmit every 60 seconds. Although NWLink does not itself broadcast SAP messages, it and NT Server, if they must communicate with pure IPX environments, may have to deal with the additional traffic SAP broadcasts generate.

DHCP BROADCASTS.   As we mentioned earlier in this chapter, DHCP does simplify IP client configuration. But the protocol's presence on an NT server, or on any server, both complicates the configuration, maintenance, and monitoring of that server, and increases network traffic. That's because DHCP broadcasts in order to do its job of assigning client IP addresses on the fly.

WINS REPLICATION.   Windows NT Server offers a mechanism for reducing broadcast traffic generated by the need to determine and resolve node addresses. That mechanism is the *Windows Internet Naming Service* or WINS, which functions in this way.

Be aware, though, that if you configure multiple WINS servers, as Windows NT allows you to do and Fig. 8-5 illustrates, the presence of these multiple servers and their need to communicate itself engenders

Figure 8-5
This is the jumping-
off point for configur-
ing WINS under NT
Server 4.

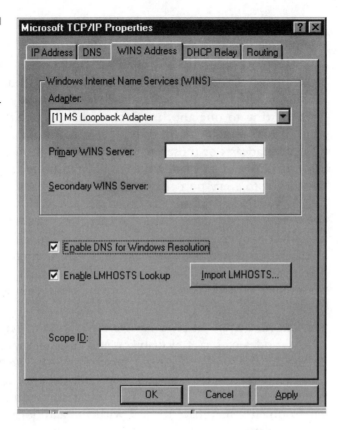

excess network traffic. However, if WINS is properly and tightly configured, this excess can be minimized.

Performance. At least some of Microsoft's networking experts believe that the performance, that is the efficiency, of protocols depends as much upon the efficiency and configuration of protocol drivers as it does upon the protocols themselves. Considering the complexity of the architecture of such protocol suites as TCP/IP, and the intricacy of the operating system networking subsets that manipulate them, such a position may be something of an overgeneralization. However, this assumption must nonetheless significantly affect protocol performance under NT, since that OS's drivers were designed with this paradigm in mind. So, in this section, we review the effects on protocol performance of NT's transport drivers.

NBF, designed to support NetBIOS and at its best in small LANs, is very fast in such environments. However, for the same reasons, it performs poorly across WANs or in Internet connections. NT's TCP/IP implementation straddles the performance fence, so to speak, being slightly slower than NBF in small LANs, but slightly faster on WAN or Internet connections. NWLink is the slowest NT driver in either environment.

Routing. Windows NT requires that you explicitly configure its multiprotocol routing capabilities if it is to be able to recognize or use the TCP/IP *routing information protocol* (RIP). If you've done so, NT can act as a static TCP/IP router, meaning that it can direct network traffic through a predefined and unchanging set of paths. However, if your environment's traffic patterns and volumes are such that:

- bottlenecks arise and must be avoided
- a high degree of communication between heterogeneous segments must take place
- paths must be ensured for critical traffic

in other words, if your network requires dynamic routing, you're going to have to fall back on third-party routers and their RIP implementations to route TCP/IP traffic. Of course, by introducing these additional devices into your NT-based network, you'll introduce at least some bandwidth overhead as well.

The same is true of the Windows NT IPX/SPX implementation, but for different reasons. NWLink relies on RIP over IPX for route and router discovery services. When NWLink loads, it sends out a request to the NetWare servers with which it is trying to converse for a network number, which it then uses for addressing at the IPX level.

Remote Access Server

Within Windows NT, connections from remote clients are handled by the operating system's Remote Access Server. In this section, we scrutinize three of the scenarios that the server must frequently manage.

1. A client sending a message
2. A LAN sending a burst of multicast traffic
3. A LAN sending a single datagram to a single client

A Client Sends a Message

Before it will handle any transmissions from remote clients, RAS checks to see if its NetBIOS Gateway function is enabled, the default being that it is. In this case, RAS forwards all messages from remote clients to all segments and all transport drivers. Needless to say, this practice is not the most efficient in terms of bandwidth consumption.

A LAN Sends a Burst of Multicast Traffic

A RAS server that receives multicast traffic handles it in this way:

First, RAS determines whether broadcasts are enabled; its default is that they are disabled. If RAS is indeed operating under this default, it will discard broadcast, but not multicast packets in the burst. Then it will check its *multicast forwarding rate,* an integer that represents the number of seconds after which RAS will forward multicast traffic to remote clients. Finally, RAS looks at its `DisableMcastFwdWhenSessionTraffic` internal parameter, which, if set to 1, causes RAS to assume that:

- the intended client is busy with some other task
- it must therefore drop the multicast packets

If:

- broadcasts are enabled
- the multicast forwarding rate has been determined
- RAS is confident that the intended destination is not busy

the RAS Server will forward multicast traffic to remote clients in this way.

- First, the RAS server will check the internal parameter `Check MaxDgBufferedPerGroupName`, whose default value is 10. RAS does this because it knows what we learned in Chap. 5, that intra-LAN and asynchronous communication speeds differ greatly, and that, therefore, traffic moving from a LAN to such lines must be buffered. Here's where `MaxDgBufferedPerGroup Name` comes in. This parameter represents the maximum number of datagrams that a RAS server will buffer for every multicast group. So, when this parameter is set to its default of 10, *every 11th datagram will be dropped,* up to a maximum of 255.

■ Next, the RAS server checks to see which clients are associated with the group name for whom the multicast is intended.

■ Finally, RAS sends the multicast traffic to each of these clients.

NT Server's RAS operates as one might expect, and efficiently, in checking for and forwarding multicast traffic only to remote clients associated with groups that have signed up to receive such traffic. However, a RAS server's practice of discarding as many as 255 packets might contribute to network congestion, depending upon whether and how often the server later tries to retransmit those packets.

A LAN Sends a Single Datagram to a Single Client

As we noted in the preceding section, a RAS server buffers all traffic it is to transmit. No internal parameter exists through which this initial buffering can be controlled. Rather, RAS waits until it detects a request from a remote client to receive network traffic. Upon noticing such a request, the RAS server forwards the datagram from its buffer to the requesting, remote client.

Three things are significant about this technique. First, the speed with which RAS retrieves the packet from its buffer and sends it on its way to the requesting client can of course be affected by other, concurrent demands for NT Server resources. Second, the RAS server's reliance upon requests from clients is a method somewhat analogous to the UNIX networking subsystem's being interrupt-driven, and can incur similar delays for similar reasons, such as those engendered by differing process priorities. Finally, pending transmissions to groups have a lower forwarding priority than immediate or pending transmissions to individual requesting clients. As a result, delays in group, that is multicast, transfers may accumulate.

Windows 95

Let's get the most significant limitation of Windows 95 as a networking platform out of the way right now. This OS can't be used as a dial-in server for any clients other than those running NetBEUI.

In view of this, we examine, in the remainder of this section, the behavior and effect on bandwidth of Windows 95 clients to a Windows NT server.

Windows 95 and PPTP

Windows 95 clients (and Windows NT Workstation clients as well) have had available, since late 1996, the *point-to-point tunneling protocol,* or PPTP. This protocol, developed by Microsoft and submitted by them to the *Internet Engineering Task Force* (IETF) for inclusion in the IP standard, is actually a means of encryption that:

- carries out user authentication
- can carry both NetBEUI and IPX traffic by transmitting these over TCP/IP messages

As such, PPTP is significant for a number of reasons, including:

- its ability to work around the routing limitations of NetBEUI and NWLink
- its ability to function within an environment or to handle remote traffic
- most significantly in the context of bandwidth, and simply of availability, its requiring all nodes it is to service to have access to a PPTP-enabled router

PPTP accomplishes either intra-LAN or remote connections in the same way. When a PPTP connection is established, protocols on the client react as if a dial-up connection had become active. PPTP itself, though, requires only that:

- some adapter in the client be active
- that adapter, be it NIC or modem, be bound to and running TCP/IP

PPTP needs the latter because it uses TCP/IP to tunnel packets.

Limitations to TCP/IP-PPTP Routing

When a PPTP-based client, whether dial-up or LAN, must connect to a host running TCP/IP, that client may experience some loss of service.

The root cause for such problems lies in PPTP's understanding of TCP/IP *default gateway routing.*

Such routing functions in this way.

- Determine if a requested destination is on the local network. If it isn't...
- Check to see if the destination is specified in a routing table. If that's not the case either, ...
- Forward the traffic to a predefined default gateway router, and assume that this router will further forward the traffic correctly.

Default gateway routing works well if a single machine is connecting to a host through a single piece of hardware. However, in situations like those in which a node has a NIC connection to its local LAN and a modem through which it may dial up remote networks, PPTP-TCP/IP routing as accomplished by Windows 95 can break down. For instance, if you:

- dial up an ISP
- then make another connection, perhaps to your LAN's PPTP server

Windows 95 will react by:

- considering the first connection to be through the default gateway
- replacing the identity of that gateway with another that represents the gateway through which you made the second connection
- thereby denying you access to machines that were reachable through the first gateway

Since such now-inaccessible machines can include DNS or WINS servers, traffic from your client may never reach its destination, because destination names and addresses won't be correctly handled, or handled at all.

This incompatibility of TCP/IP default gateway routing, which was intended to work with a single network, and PPTP connections, which may be established either by means of a dial-up or a local link, which can result in an incorrectly identified default route, can be resolved only in one way. When the more recent PPTP connection is released, all the connection to the first destination is restored.

IN A NUTSHELL

1. Windows NT Server's combination of

   - support for multiple protocols

   - use of the NetBIOS application interface

   - employing the Microsoft Transport Driver Interface to map protocol addressing to NetBIOS

   introduces a level of complexity to network traffic not found under other operating systems.

2. The Windows NT TCP/IP driver uses a NetBIOS compatibility layer over TCP/IP itself.

3. The Windows NT IPX driver uses a NetBIOS compatibility layer over IPX itself.

4. Microsoft recommends that Windows NT servers rely on the dynamic host configuration protocol to assign IP addresses, despite the overhead inherent in using DHCP.

5. Windows NT Server, because it can maintain only static routing tables

   - must rely on third-party routers for dynamic routing

   - can thereby incur the bandwidth overhead typical of such routers' broadcasts to one another

6. Windows 95, when relying on the point-to-point tunneling protocol, may experience bandwidth overhead caused by that protocol's:

   - sometimes doubling up routable and poorly or nonroutable protocols, for example, carrying NetBEUI messages on the back of TCP/IP packets

   - practice of making connections reached through an initial default gateway invisible when a client simultaneously connects to a second such gateway

LOOKING AHEAD

In Chap. 9, we leave operating system constraints on bandwidth, and turn instead to limitations imposed by what must be, in today's networking world, the most widely used network application: the Web server.

Apache, Internet Information Server, and Demands on Bandwidth

Chapter 9 scrutinizes the effects on overall bandwidth consumption of two examples of a type of application certain to have a significant impact on that consumption—Web servers. The examples we investigate are:

- Apache
- Internet Information Server

Apache

When Apache 1.3 is downloaded, the distribution, as Fig. 9-1 describes, includes source code for a number of modules.

These include:

- `core`: core Apache features
- `mod_access`: host-based access control
- `mod_actions` (Apache 1.1 and later): filetype/method-based script execution
- `mod_alias`: aliases and redirects
- `Mod_asis`: as-is file handler
- `mod_auth`: user authentication using text files
- `mod_auth_anon`: anonymous user authentication, FTP-style
- `mod_auth_db`: user authentication using Berkeley-style databases
- `mod_auth_dbm`: user authentication using DBM files
- `mod_autoindex`: automatic directory listings
- `mod_browser` (Apache 1.2.* only): sets environment variables based on User-Agent strings; replaced by `mod_setenvif` in Apache 1.3 and up

Figure 9-1
As you can see, Apache offers quite a list of downloadable code, and documentation as well.

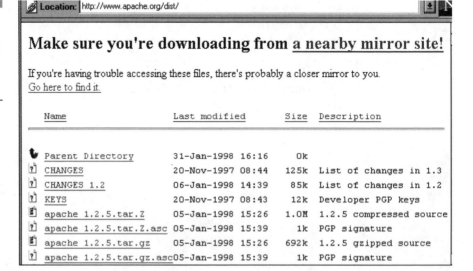

Location: http://www.apache.org/dist/

Make sure you're downloading from a nearby mirror site!

If you're having trouble accessing these files, there's probably a closer mirror to you.
Go here to find it.

Name	Last modified	Size	Description
Parent Directory	31-Jan-1998 16:16	0k	
CHANGES	20-Nov-1997 08:44	125k	List of changes in 1.3
CHANGES 1.2	06-Jan-1998 14:39	85k	List of changes in 1.2
KEYS	20-Nov-1997 08:43	12k	Developer PGP keys
apache 1.2.5.tar.Z	05-Jan-1998 15:26	1.0M	1.2.5 compressed source
apache 1.2.5.tar.Z.asc	05-Jan-1998 15:39	1k	PGP signature
apache 1.2.5.tar.gz	05-Jan-1998 15:26	692k	1.2.5 gzipped source
apache 1.2.5.tar.gz.asc	05-Jan-1998 15:39	1k	PGP signature

- `mod_cern_meta`: support for HTTP header metafiles
- `mod_cgi`: invoking CGI scripts
- `mod_cookies` (up to Apache 1.1.1): support for Netscape-like cookies; replaced in Apache 1.2 by `mod_usertrack`
- `mod_digest`: MD5 authentication
- `mod_dir`: basic directory handling
- `mod_dld`: start-time linking with the GNU libdld
- `mod_dll`: start-time module linking with Win32 DLLs
- `mod_env`: passing environment variables to CGI scripts
- `mod_example`: demonstrates the API for Apache 1.2 and up
- `mod_expires` (Apache 1.2 and up): apply expiration headers to resources
- `mod_headers` (Apache 1.2 and up): add arbitrary HTTP headers to resources
- `mod_imap`: image map file handler
- `mod_include`: server-parsed documents
- `mod_info`: server configuration information
- `mod_isapi`: Windows ISAPI support
- `mod_log_agen`: logs user agents.
- `mod_log_common` (up to Apache 1.1.1): logging in the Common Logfile Format; replaced by mod_log_config in Apache 1.2 and up
- `mod_log_config`: user-configurable logging
- `mod_log_referer`: logs document references
- `mod_mime`: determines MIME type
- `mod_mime_magic`: determines MIME type by using magic numbers
- `mod_negotiation`: content negotiation
- `mod_proxy`: caching proxy abilities
- `mod_rewrite` (Apache 1.2 and up): URI-to-filename mapping using regular expressions
- `mod_setenvif` (Apache 1.3 and up): sets environment variables based on client information

- `mod_speling` (**Apache 1.3 and up**): automatically correct minor typos in URLs
- `mod_status`: **server status**
- `mod_userdir`: **user home directories**
- `mod_unique_id` (**Apache 1.3 and up**): generates unique request identifier for every request
- `mod_usertrack` (**Apache 1.2 and up**): user tracking using Cookies (replacement for `mod_cookies.c`)

The members of this extensive list may be compiled into an Apache executable, in a way like that outlined in Fig. 9-2.

As the number and nature of its modules indicate, Apache was designed to be correct first, and fast second. Nevertheless, this most widely used Web server performs quite acceptably in most environments. In sites that use less than 10 Mbits of outgoing bandwidth, Apache functions well even on a low-end Pentium platform. In sites that require more outgoing bandwidth, it's often the case that more than one server is present as well, handling such Web-related operations as database transactions. In such environments, calculations and evaluations of the speed and efficiency of a Web server become more complicated. Further complicating such appraisals is the tendency of many people, infor-

Figure 9-2
Compiling Apache
configures it as well.

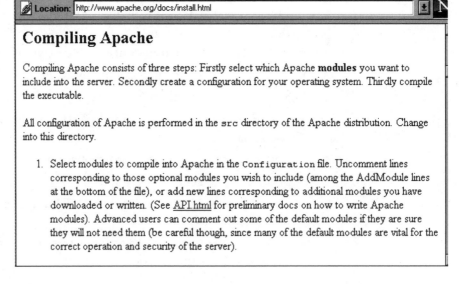

Location: http://www.apache.org/docs/install.html

Compiling Apache

Compiling Apache consists of three steps: Firstly select which Apache **modules** you want to include into the server. Secondly create a configuration for your operating system. Thirdly compile the executable.

All configuration of Apache is performed in the `src` directory of the Apache distribution. Change into this directory.

1. Select modules to compile into Apache in the `Configuration` file. Uncomment lines corresponding to those optional modules you wish to include (among the AddModule lines at the bottom of the file), or add new lines corresponding to additional modules you have downloaded or written. (See API.html for preliminary docs on how to write Apache modules). Advanced users can comment out some of the default modules if they are sure they will not need them (be careful though, since many of the default modules are vital for the correct operation and security of the server).

mation professionals as well as users, to look largely or only at raw performance numbers in judging the efficiency of a Web server.

Apache's designers operate from the premise that although there is indeed a standard for minimum acceptable throughput performance for a Web server, speed of transmission beyond that minimum is considered critical only by a very small percentage of Web sites. Nonetheless, in order to make Apache acceptable even in such markets, the server's designers made a concerted effort, in implementing Apache 1.3, to make their product fully competitive with other high-end Web servers in terms of bandwidth efficiency and transmission speed.

What's more, many of those involved in developing, enhancing, and maintaining Apache admit to, as one member of the Apache Group puts it, "just plain want[ing] to see how fast something can go." As a result, the Apache Group has developed guidelines for, again in their words, "squeeze[ing] every last bit of performance out of Apache's current model, and . . . understand[ing] why it does some things which slow it down."

NOTE: *The Apache Group's efforts to fine-tune the server assume running Apache 1.3 on UNIX. Only some of their enhancement efforts apply to Apache on NT, since the server was only ported to the latter operating system within the past year.*

Hardware and Operating System Issues

The Apache Group's observations and recommendations regarding fine-tuning their server are a gold mine of down-to-earth advice. And we should point out that many of their suggestions can be valuable, with appropriate translation to the syntax and configuration of particular environments, on a variety of hardware/operating system platforms. We've summarized their thoughts in this section.

The Role of RAM.   To the Apache Group, the single biggest hardware issue affecting Web server performance is RAM. In their words, "a webserver should never ever have to swap, [since] swapping increases the latency of each request beyond a point that users consider 'fast enough.' This causes users to hit stop and reload, further increasing the load."

As a means of minimizing the likelihood of this scenario, the group recommends configuring the server with a close eye to keeping the maximum number of client sessions, defined in Apache by the parameter `MaxClients`, to a value that allows the server to avoid excessive spawning of child processes, thereby helping it to avoid swapping.

━━ ━━ ━━ ━━ ━━ ━━ ━━ ━━ ━━ ━━ ━━ ━━ ━━ ━━ ━━ ━━ ━━ ━━

NOTE: *Apache's MaxClients directive, like other configuration directives we mention in this section, is always available to the Apache administrator. Max-Clients can be summarized as follows.*

```
Syntax: MaxClients number
Default Value: 256
```

The MaxClients directive sets the limit on the number of simultaneous requests that can be supported. Whatever value you supply for this parameter will preclude Apache from creating any number of child (server) processes in excess of that value.

Beyond that the rest is mundane: get a fast enough CPU, a fast enough network card, and fast enough disks, where *fast enough* is something that needs to be determined by experimentation.

The Role of the Operating System. The Apache Group considers the choice of operating system platform for the server to be something that must be defined primarily by the needs of particular environments. But the group does make one recommendation regarding this choice. They stress that the latest operating system TCP/IP patches should always be applied. This advice rests on shortcomings of the UNIX networking subsystem like those we discovered in Chap. 7. One Apache group member summarizes such shortcomings by saying, "HTTP serving completely breaks many of the assumptions built into UNIX kernels up through 1994 and even 1995."

Run-Time Configuration Issues

Again, we'd like to emphasize that, even though the points discussed and the options suggested in this section pertain to Apache running under some flavor of UNIX, they're down-to-earth enough to be applicable to a number of other environments as well.

HostnameLookups. Prior to Apache 1.3, the default setting for the parameter `HostnameLookups`, which represents whether or not the server will carry out a domain name server lookup for client requests, was On. This default enabling of DNS lookups for every request increased latency for each, since the DNS lookup had to complete before the request would be satisfied.

As a result, in the fine-tuning that went into Apache 1.3, this setting was made to default to Off. However, even the most recent versions of the server may still experience delays, if Apache has been built to include directives that allow or deny traffic from specific domains. Such directives can engender two, not just one, DNS lookups, the second being carried out in order to try to ensure that the first wasn't the victim of a spoof. As an alternative to these double DNS lookups and to help accomplish optimum performance, the group suggests not only avoiding such directives but relying on IP addresses rather than domain names.

━━ ━━ ━━ ━━ ━━ ━━ ━━ ━━ ━━ ━━ ━━ ━━ ━━ ━━ ━━ ━━

NOTE: *Apache's HostNameLookups configuration directive, which can be summarized as follows:*

```
Syntax: HostNameLookups on | off | double
Default Value: HostNameLookups off
```

enables DNS lookups so that host names can be logged and passed to CGIs. The value double refers to doing double-reverse DNS lookups as described above, a practice that in TCP/IP programming parlance is known as PARANOID.

However HostnameLookup is defined, when the Apache module `mod_access` *controls access by hostname, a double reverse lookup will be done.*

Another directive- and DNS-related constraint on lookups exists that can actually improve throughput. It is possible to define the scope of these directives in such a way as to cause DNS lookups to be carried out only on requests that match the scope's definition. For example, this code fragment would disable DNS lookups except for those required by requests for .html or .cgi files.

```
HostnameLookups off
#Turn off host name lookups . . .
<Files ~ "\.(html|cgi)$>
# . . . but then, define an area of the file subsystem . . .
HostnameLookups on
# . . . for which requests for files will be permitted to carry out
such lookups
</Files>
```

Another alternative recommendation for dealing with the overhead that host lookups can generate is to make calls to modules like Apache's `gethostbyname` only from the specific CGI programs that need those names.

FollowSymLinks and SymLinksIfOwnerMatch. The nuances of a number of Apache's operating characteristics can be controlled through the server's configuration directive Options. We've summarized this directive below.

```
Syntax: Options [+|-]option-name [+|-]option-name . . .
```

The Options directive controls which Apache features are available for a particular directory.

You make no such features available, by setting option-name None. Setting this parameter to All, on the other hand, makes all Apache's special features except MultiViews available in a given directory. Other values that can be supplied for option-name are outlined in Table 9-1.

Any options preceded by a + are added to the options currently in force for a directory, whereas any preceded by a - are removed from the set currently in force.

TABLE 9-1

Defining Apache Features for Specific Directories

This Option	Allows a Directory
ExecCGI	To execute CGI scripts
FollowSymLinks	To have the server follow symbolic links in the directory, without the server's changing the pathname used to match against <Directory> definitions
Includes	To use server-side includes
IncludesNOEXEC	To use server-side includes, but to disable the #exec and #include commands in CGI scripts
Indexes	To be represented as a formatted listing by the server, if a URL that maps to the directory is requested, and no DirectoryIndex, that is, no file such as index.html, exists in the directory
SymLinksIfOwnerMatch	To present symbolic links that the server will, only if the target file or directory is owned by the same user id as the link

Consider the following Apache configuration example.

```
<Directory /web/docs>
Options Indexes FollowSymLinks
</Directory>
<Directory /web/docs/spec>
Options Includes
</Directory>
```

Given this set of definitions, Includes will be allowed for the directory /web/docs/spec, whereas indexes and the ability to follow links will exist for the /web/docs directory.

How does all this relate to bandwidth consumption? In advising about how to define the features available to your content subsystem, the Apache Group makes this caution. If you do not assign the ability to follow links by issuing the directive

```
Options FollowSymLinks
```

or if you do use, for that purpose, the directive

```
Options SymLinksIfOwnerMatch
```

Apache will make one extra system call for every filename component, in order to check links. So, if you used code like:

```
DocumentRoot /www/htdocs
<Directory />
Options SymLinksIfOwnerMatch
</Directory>
```

and a request is made for the URI

```
/index.html
```

Apache will check the status of these files and paths:

- /www
- /www/htdocs
- /www/htdocs/index.html

Apache doesn't cache the results of these checks for the status of links. Rather, such tests must take place for every client request. Since this is the case, and if your environment is one in which the added security offered by checking links is important, your Apache configuration would have to include a directory-by-directory use of not only the `Fol-lowSymLinks` parameter, but the more secure `SymLinksIfOwner` parameter as well, as in this example:

```
DocumentRoot /www/htdocs
<Directory />
Options FollowSymLinks
</Directory>
<Directory /www/htdocs>
Options -FollowSymLinks  + SymLinksIfOwnerMatch
</Directory>
```

But, to ensure best performance, albeit at the price of higher security, the Apache group recommends setting only the `FollowSymLinks` parameter, but setting it for every document path.

Negotiation.   In discussing getting the most out of Apache, its documentation makes a point that can be applied to any Web server. *If at all possible, avoid content negotiation.* To accomplish savings in bandwidth in this way, without completely sacrificing the ability to choose from a set of content, the Apache Group suggests using a complete list of content options, such as:

```
DirectoryIndex index.cgi index.pl index.shtml index.html
```

rather than a wildcarded list like:

```
DirectoryIndex index*
```

If, in your explicit list of content choices, you place the items in descending order of frequency of selection, you can offer the best possible compromise between content negotiation and bandwidth savings.

Compile-Time Configuration Issues

Unlike most commercial Web servers, Apache provides source code that may be modified and recompiled to tailor the server to meet environment-specific needs. It is at the point of compilation that the administrator chooses those operating parameters he or she wishes to include in the Apache executable. This being the case, the specifics of what we're about to discuss will differ in the context of commercial Web servers. The context of the points in this section, though, can apply to any such application.

mod_status As we noted early in this section on Apache, `mod_status` is a core Apache feature that, when included in a build of the server, provides server status displays as Apache runs. If, when Apache is compiled, you include `mod_status` and at the same time set the compile-time parameter

```
Rule STATUS=yes
```

the resulting Apache executable will do duplicate system calls to the UNIX time-of-day utility, in order to make timing information available to Apache status reporting. Such duplication, of course, draws off memory and CPU cycles that might otherwise have been available directly to transmission tasks. So, under Apache, if you seek the most efficient use of bandwidth, you must set

```
Rule STATUS = no
```

Under other Web servers, you could, as an analog, configure the server's status reporting to be as lean as possible.

Accept Serialization—Multiple Sockets. One shortcoming that UNIX-based Web servers may experience results from a shortcoming in the operating system's socket API. That API encourages the use of multiple code statements to listen on multiple ports or addresses. Apache's model includes the possibility of multiple child processes being spawned by a single client request, and the further possibility that all idle child processes might test a socket for new connection requests at the same time. This combination in turn produces the opportunity for:

- children to block one another
- blocked children all to attempt to respond to the next available open socket
- since only one can succeed, the rest of the children again blocking one another. . .

and so on until enough new requests have appeared on a given socket to satisfy all children.

One solution is to make sockets *nonblocking.* In this case, the accept of a connection won't block other, waiting child processes. Rather, they will go on their way, so to speak, immediately. But such an answer wastes CPU time. Suppose you have 10 idle children that want to select a connection. Suppose further only one connection arrives. In this scenario, even after the connection is accepted, the remaining nine children will wake up, try to accept the connection, fail, and loop back into select mode; therefore, nothing has been accomplished. Nor have any of these nine children been able to process requests that were placed on other sockets until they return to select mode. In most environments, such a solution can't be called productive.

Another solution, the one used by Apache, is to serialize the order in which child processes can return to select mode. Apache does this by using a mutually exclusive semaphore that can be held by only one child at a time. In this way, the server avoids having several children unavailable to processing other sockets. However, this answer has a decided drawback. It simply doesn't work under some versions of UNIX, like IRIX, which run on multiprocessor machines.

The Pre-Forking Model

Under UNIX, Apache uses a pre-forking model. The server does not itself serve any requests or service any network sockets. Rather, it simply forks, creating child processes that do either. These child processes serve multiple connections, but serve them one at a time, before dying. The parent, that is, Apache, spawns new or kills off old children in response to changes in the load on the server. In order to accomplish this, Apache monitors a status file that the children maintain.

This model for servers offers a robustness that other models do not. In particular, code for the parent process is very simple, and offers a high degree of confidence that the parent process will do its job without error. Pre-forking also makes Apache very portable across flavors of UNIX, historically and still an important goal for the server.

However, the pre-forking model does have performance drawbacks. Most important to our discussion are these:

- the memory and processing overhead inherent in forking processes
- the overhead inherent in switching between child processes
- the memory overhead inherent in multiple processes
- the fewer opportunities for data-caching between requests inherent in forking

In attempting to address the limitations of a forking model, the Apache group points out that:

- the server's core code is already multithread aware
- version 1.3 is multithreaded on NT
- there are two experimental implementations of threaded Apache
- redesign for version 2.0 of Apache will include abstractions of the server model that will allow the support of both the pre-forking model and threaded models

Internet Information Server 4.0

This most recent incarnation of Microsoft's Windows NT–based Web server has a total of 19 significant features. These are:

- Active Directory Service Interface: a single interface for users to multiple directory systems
- Active Server pages: sample ActiveX pages for server code
- Authentication Server: using the Remote Authentication Dial-in User Services, or RADIUS, protocol to check remote user-connection attempts
- ActiveX: objects for data access
- FrontPage Server extensions: making FrontPage available as a server Web publishing tool
- Index Server: to index not only Web but other files
- Jscript: complete with user's guide; Microsoft's 100 percent JavaScript-compliant language
- Mail Server: to manage email with the simple mail transfer protocol (SMTP)
- Microsoft Management Console: a manager module for both NT Server itself, and for its Internet services
- News Server: to manage news groups
- ODBC: the Open Database Connectivity system for administering remote databases
- Posting Acceptor: which posts Web pages

- Script Debugger: a tool for debugging Jscript, Java, Perl, VBScript, and other code

- Site Analyst: to create and maintain maps of Web sites

- Transaction Server: to create, configure, and distribute server-based applications

- Usage Analyst: to create and maintain Internet Information Server logs

- VBScript: with a complete user guide

- Web Publishing Wizard: to accomplish automatic publishing of Web pages

- Windows Scripting Host: to manage both Web-related and general server scripts

The mere presence of so many tools constitutes a potential drain on server resources, particularly on RAM. When you consider that all these tools, like IIS and its underlying operating system Windows NT, are almost entirely GUI-based, the extent of that potential becomes more clear. However, beyond these generalities, there are particular IIS 4 features that, through their relationships to data transfer, affect bandwidth consumption more directly. We investigate these features now.

Active Directory Service Interface (ASDI)

ASDI is made up of a number of modules that provide routing and routing support for clients that must communicate with, for example, such differing directory services as NetWare 3.x, NDS, and Windows NT. Such routing implies protocol translation, not only a drain on server RAM but also a source of delay in throughput.

Internet Authentication Server (IAS)

This server, once part of Windows NT Server proper, handles not only user authentication and authorization, but also user accounting. IAS relies on the RADIUS protocol, but handles access requests locally, rather than remotely as RADIUS ordinarily would. Therefore, in effect IAS can be considered a server for an authentication service. It presents the potential for bandwidth overhead to IIS environments by adding this extra layer of operational complexity, as well as by adding another protocol to Windows NT's communications kitbag.

Internet Mail Server and Internet News Server

These two IIS modules, whose default news groups we've shown in Fig. 9-3, and some of whose configuration Figs. 9-4 and 9-5 illustrate, rely on application-level TCP/IP protocols. Mail Server uses a version of SMTP tuned to heavy traffic.

News Server rests on the *network news transport protocol* (NNTP), which, among other things, provides a common client/server command set through which news groups may be accessed.

In addition, both these servers support secure communications, through the *secure sockets layer* or SSL group of protocols. Such additional functionality, although valuable, nonetheless can contribute to band-

Figure 9-3

Out of the box, this installation of News Server offers a number of news groups.

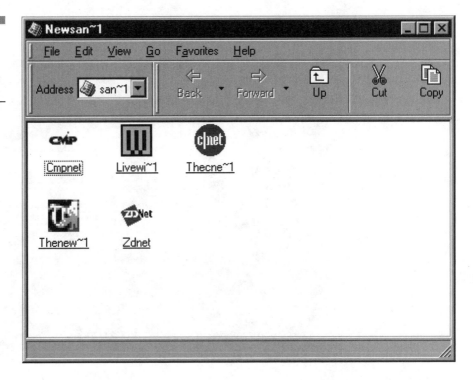

Figure 9-4
Configuring both
internal and Internet
mail starts here.

Figure 9-5
Post offices can be
password-protected.

width overhead, through the presence of these extra protocols, and through their need to communicate with other protocols.

IN A NUTSHELL

1. Apache's performance can be adjusted by configuring it at compilation to include or exclude performance-related modules, and by setting certain of its configuration directives to optimize performance.

2. Among the Apache modules that are most likely to affect throughput are: `mod_access`, `mod_alias`, `mod_auth`, `mod_auth_anon`, `mod_cern_meta`, `mod_imap`, `mod_include`, `mod_mime`, `mod_negotiation`, `mod_proxy`, and `mod_rewrite`.

3. Among the configuration directives whose settings can affect Apache's resource and bandwidth consumption are `MaxClients`, `HostnameLookups`, and `FollowSymLinks`.

4. By introducing additional protocols and the need for additional interprotocol conversations, Internet Information Server also introduces the potential, in its ADSI, IAS, IMS, and INS modules, for bandwidth overhead.

LOOKING AHEAD

Chapter 10 examines a model for the use of memory by distributed applications. By studying this model, we can discover the demands such applications can place on bandwidth, through the demands they place on the server resource most critical to bandwidth efficiency.

Client Applications and Demands on Bandwidth: A Paradigm

Chapter 10 probes the effects on bandwidth consumption of the use of memory by distributed applications.

Distributed Memory

Many if not most *remote database management systems* or RDBMS rely on the concept of transactions. *Transactions* break down what the user might think of as a single process, such as modifying a database record, into a number of processes, such as:

■ determining the user's access permissions for the database and specific record

■ determining if the record in question is locked or available

■ retrieving the record

■ making the change

■ committing the transaction, that is, saving the changed record to disk

Transactions introduce greater stability into database manipulation, by offering such characteristics as *atomicity* and *recoverability.* However, relying on transactions has its downside. The performance of transaction-based systems can be limited by the performance of the disks that house its data.

In April 1997, researchers at the Foundation for Research and Technology (FORTH) in Hellas, Crete, announced the results of investigations into using the collective main memory in a network of workstations to improve the performance of transaction-based systems. This imaginative approach shares many of the characteristics of the effect in general of distributed applications on server resource use. Clearly, this model also can be considered an example of the effects of such applications on bandwidth efficiency, particularly since the use of server RAM impacts that efficiency. For these reasons, we begin our discussion of the effects of distributed applications on bandwidth with a closer look at the *distributed memory model.*

The FORTH researchers cite two areas where remote memory can improve the performance of a transaction-based system.

Read Access

The collective RAM of a network of workstations can be used as a large cache for a transaction-based system, as Fig. 10-1 illustrates.

Such a shared cache, obviously much larger than that any one workstation could provide, can therefore hold much larger amounts of data.

Figure 10-1
Shared caching can
significantly increase
overall cache size
and availability.

Figure 10-1
Shared caching can
significantly increase
overall cache size
and availability.

By using memory from a number of nodes, caching can be made more efficient.

Reading data from remote memory caches proved, in the Hellas experiments, significantly faster than reading data from a local drive. But such reads relied on high-speed networks to provide higher throughput and lower latency than high-speed disks. Where such transmission paths aren't available, the trend cited in the Hellas study of the performance disparity between disks and network caches increasing over time may not manifest.

Write Operations

In order to provide for a transaction-based system's being able to recover to an orderly state after a hardware or a software failure, such systems often use synchronous write operations to force all modified data to disk at transaction commit time. The FORTH researchers feel that the distributed RAM provided by a network of workstations has reliability comparable to that of disks, and therefore can be used to cache data against the possibility of a system crash.

The factor most strongly influencing the efficiency of synchronous write operations is the *latency* of the medium to which the write is done. The latency of remote memory, although dictated by overall network latency, can be significantly lower than disk latency. Further, remote memory latency, largely generated by protocol processing, can improve if processor speed or the amount of RAM improves. Disk latency, on the other hand, controlled largely as it is by mechanics such as the movement of the read/write head of a drive, cannot be as easily improved upon.

The Effects of Network Architectures

Recent trends in the evolution of network media make the concept of distributed memory caching to speed up synchronous disk I/O operations more attractive, because the latency of these media has decreased noticeably over the last few years. For example, Ethernet and FDDI have latency in the range of several hundred microseconds. ATM networks have latency in the range of a few hundred microseconds. More recent network media have latency in the range of a few microseconds.

Based on such trends, the researchers at FORTH feel that transaction-based systems can effectively use remote RAM to avoid the delays inherent in data transfers to and from disks. To demonstrate this approach, those researchers implemented it within two existing transaction-based systems: *The EXODUS storage manager* and the *recoverable virtual memory* or RVM System.

The Experimental Design

During its lifetime, a transaction may:

- make a number of disk accesses to read its data
- massage that data in some way several times
- write results to the DBMS equivalent of a log file
- only then, if everything has succeeded to this point, commit its processing to disk

The number and latency of read operations may be reduced through the allocation of larger-than-usual amounts of server memory. But it's more difficult to pare down the number of disk write operations at transaction commit time, because a transaction's modified data and meta-data have to reach stable storage before the transaction commits. Current transaction-based systems use disk as the stable storage medium, since disks usually survive power and software failures. The FORTH researchers feel that distributed memory can be made similarly reliable.

Figure 10-2
As you can see, remote memory is only one of many areas being investigated at FORTH.

ICS - FORTH

Computer Architecture and VLSI Systems Group

The *Computer Architecture and VLSI Systems Group (AVG)* is one of the five research and development groups of the Institute of Computer Science (ICS) of FORTH. AVG conducts research and development in the architecture and design of computer and communication systems, and their implementation, at the IC, board, and system level.

Recent research topics include **Networks of Workstations (NOWs), ATM networks, high-speed ATM switching, and operating systems for NOWs.**

- **General Description (.ps) or (.ps.gz) or single-page version.**

NOTE: *For a complete discussion of the mechanisms through which this can be accomplished, visit the URL illustrated in Fig. 10-2:* http://www.ics.forth.gr/proj/arch-vlsi. *Essentially, the FORTH method relies on mirroring and uninterruptable power supplies (UPSes) to make distributed memory as reliable as disks.*

You might also want to glance at the activities of the FORTH networking group, at http://www.ics.forth.gr/ICS/acti/netgroup. *As Fig. 10-3 shows, this group is conducting research into many areas which affect bandwidth efficiency.*

Figure 10-3
Evaluating the performance of your network might be aided by reviewing the ICS Networking Group's research.

Ομάδα Τηλεπικοινωνιών και Δικτύων

Telecommunications & Networks Group

Telecommunications and Networks Division

Members

Contacts

Funded Research

Documentation

The Telecommunications and Networks Division is involved in theoretical and applied research in the area of high-performance communication networks. The theoretical work concentrates on topics such as the use of applied probability models for the evaluation of the performance and the *Quality-of-Service (QoS)* management of *ATM* networks, the theory of pricing network and user services, and the development of formal methodologies and tools for the specification and verification of computer-communication protocols and real-time systems.

The Experimental Design

In this section, we take a closer look at the experimental design that underlies FORTH research into distributed memory.

EXODUS and RVM

To illustrate the benefits of remote memory, the researchers in Hellas modified both the lightweight transaction-based system RVM and the EXODUS storage manager to use remote memory instead of disks for synchronous write operations. Both EXODUS and RVM were changed so as to keep a copy of their log file in remote RAM as well as on disk. In the modified systems, synchronous write operations were replaced by two tasks:

1. a synchronous write to the log in the main memory of one remote workstation

2. an asynchronous write to the log on disk, performed in background to preserve a local copy of data as insurance against the possibility of remote RAM crashing

At transaction commit time, the transaction's data are synchronously written to the log in remote main memory. At the same time, these data are asynchronously written to the local magnetic disk. The transaction commits after the disk write has been scheduled, but before it completes. There is a window of vulnerability inherent in this approach. If a system crashes during this interval between scheduling and actual writing, any data still in the local memory buffer cache will be lost. However, the FORTH system can recover that data, since it also resides in remote memory. Full data loss can happen only if both local and remote systems crash during this interval.

Recovery

In the event of a workstation/network crash, our system needs to recover data and continue its operation. If the local workstation crashes, and reboots, it will read all its "seemingly lost" data from the remote memory, store them safely on the disk, and continue its operation normally. If the remote workstation crashes, the local transaction manager will realize it after a timeout period. After the timeout, the local manager may either search for another remote memory server, or just stop using remote memory, and commit transactions to disk as usual. If the network crashes, the local workstation will stop using remote memory and will commit all transactions to disk. In all circumstances, the system can recover within a few seconds, in the worst case. The reason is that at all times there exist two copies of the log data: if one copy is lost due to a crash, the system can easily switch to the other copy quickly.

Memory Service

The modified systems, called *Remote RVM* or RRVM and *Remote EXODUS* or REX are implemented completely in user space, without any operating system modifications. For each transaction manager presented by these applications, a user-level remote memory server is started on a remote workstation. This server accepts synchronous write requests from the transaction manager with which it's associated, and acknowledges those requests, so that, in the event of a transaction manager's crashing, the remote memory server reproduces the contents of the transaction manager's log file. All data involved in committed transactions is placed in memory of at least one remote workstation; that which houses the remote memory server.

Experimental Environment

The FORTH experimental environment was made up of:

- a network of eight DEC Alpha 2000 workstations running at 233 MHz, each with 128 MB RAM and a 6-GB hard drive
- connections through Ethernet, 100 Mbps FDDI, and a Memory Channel Interconnection Network, the latter a very high bandwidth network designed to handle high-performance applications

Four separate system configurations were tested in this environment.

1. the unmodified RVM system
2. RRVM running on Ethernet

3. RRVM running on FDDI

4. RRVM running on a Memory Channel Interconnection Network

I/O Block Size

The first set of experiments carried out in Hellas attempted to determine how many transactions per second the RRVM system could sustain, compared to the number of transactions per second the unmodified RVM system had proven to sustain. This experiment involved:

- a 100 MB file
- a sequence of 10,000 transactions, each of which writes a segment of the file and commits

The size of the file segment modified by each transaction, that is, of the I/O block involved, is the variable in this set of experiments. Performance of the systems tested is outlined in Table 10-1.

Log Size

The log file kept by all the examined systems and synchronously written by transactions during their commit phase can affect the systems in question in various ways. When the file increases in size past a defined threshold, RVM reads and truncates the log, and only then updates the data file.

TABLE 10-1

Performance of Remote-Memory Transaction Systems

This System	Performs Well When	But Performs Poorly When
unmodified RVM	handling no more than 40 transactions per second, for small transactions	I/O block size increases beyond 50 transactions per second
RRVM on Ethernet	handling up to 500 (small) transactions per second	I/O block size increases beyond 500 transactions per second
RRVM on FDDI	handling up to 500 (small) transactions per second	I/O block size increases beyond 500 transactions per second
RRVM on an MCIN	handling up to 3000 transactions per second	no really poor performance exhibited

Under such a technique, the log file can be viewed as a buffer between synchronous transaction writes and asynchronous data file updates.

To measure the performance effect of the size of such log files, the FORTH researchers varied, not I/O block size, but log size, while keeping I/O block size constant. All systems tested exhibited poor performance for log sizes smaller than a few KB, especially when those logs dealt with a relatively large I/O block size. Such results might seem inconsistent, but in fact they are not. Very small logs force applications to issue two synchronous write operations per transaction, the first to write data to the log at transaction commit time, and the second to empty the log to the data file. But for logs larger than 32 KB, the performance of both RRVM systems is significantly better than the performance of the unmodified RVM system, by one to two orders of magnitude.

Random Accesses

FORTH also tested the performance of all experimental transaction-based systems in a random-access rather than sequential mode. In this set of experiments, transactions accessed the data file completely randomly, and by means of an 8-MB log file.

As you might expect, all of RRVM-MC, RRVM-FDDI, and RRVM-ETHERNET performed much better than RVM under these conditions. But, in contrast to earlier experiments, the number of transactions per second sustained by RRVM-FDDI and RRVM-ETHERNET didn't as closely approach that demonstrated by RRVM-MC, primarily because of the greater number of page faults likely to occur in random, as opposed to sequential, accesses.

NOTE: *A page fault is a type of DBMS error that can take place during random accesses. Page faults entail attempts to retrieve pages that fail because of missing page boundaries.*

Distributed Memory and Network Load

A scheme like that outlined in the previous sections obviously has implications for network as well as for server performance. The studies at

Hellas suggest that on top of a lightly loaded network, RRVM performs much better than traditional RVM, without negatively affecting the performance of other distributed applications.

In order to determine:

- if the various flavors of RRVM would perform as well if several instances ran concurrently
- how the underlying, now more heavily loaded, network would handle the increased communication demands

a new experiment was designed. In this new test, several instances of RRVM clients, each with its own RRVM server, were created. All such clients ran on workstations different from those used either by the other RRVM clients, or by the specific client's RRVM server. Finally, all workstations, whether hosting client or server RRVM, were connected to the same network, either Ethernet or FDDI.

The number of client/server pairs taking part in the experiment was steadily increased, and the number of transactions per second each client/server pair handled measured. Log size and I/O block size, however, remained constant, at 8 MB and 32 bytes, respectively.

Under these conditions, the performance of RRVM-ETHERNET decreases with the number of workstations. Running on a single pair of workstations, RRVM-ETHERNET sustains close to 500 transactions per second. Running on a network that contains four RRVM client/server pairs, this system can sustain only about 150 transactions per second.

RRVM-FDDI performed consistently; however, many client/server pairs were present. Even with as many as four of these pairs, RRVM-FDDI showed no performance loss. FORTH points out that such results are to be expected, since FDDI has 10 times Ethernet's theoretical throughput, and therefore can sustain several heavily communicating stations. Nor is FDDI subject to the collisions so common to Ethernet when it operates under heavy load.

Implications

The use of *shared memory* as a distributed buffer for transaction-based systems demonstrably improves the performance of those systems. However, as a paradigm for improving the performance of distributed applications, shared memory must be viewed with caution. This model effectively doubles the number of network conversations taking place between sharing stations. If those stations are at the same time servicing other sorts of client requests, shared memory might actually cause overall performance disruptions.

IN A NUTSHELL

1. Transaction-based DBMSes epitomize many of the demands made on both server and network resources by distributed applications in general.

2. The use of distributed, shared memory to cache transactions will significantly improve the performance of transaction-based DBMSes.

3. The degree to which the performance of these transaction-based DBMSes improves with the use of shared, distributed memory caches is affected by:

   ■ the communications medium to which the sharing stations connect

   ■ the sizes of transaction log files and I/O blocks

4. While shared, distributed memory caching can improve the performance of distributed applications, it also presents the potential for negative effects on overall network throughput.

LOOKING AHEAD

In Chap. 11, we move from scrutinizing constraints upon bandwidth to examining how the effects of those constraints on transmission efficiency can be minimized. We begin Part 2 with a look at how the physical communications channel can be adjusted in service of this goal. Specifically, we examine the benefits of Fast Ethernet.

Optimizing Bandwidth Usage

11

Speeding Up
Networking Hardware

As networks grow, they also tend to change their make-up, in two important ways. First, the type of traffic they carry changes from simple file and print sharing to distributing, in a client/server environment rather than a peer-to-peer one, a much wider variety of data, such as that related to databases. Concurrently, of course, the volume of data traversing the network increases dramatically. It's at this point in their development that networks begin to experience significant congestion. That congestion can be exacerbated by the increasingly ubiquitous use of Microsoft applications, not the leanest to be found. If suites such as CAD are added to a network's profile, congestion can grow further.

Network congestion, whatever its causes, has often been compared to traffic jams in urban settings. Like a system of highways, networks tend to develop bottlenecks, spots where inordinately large amounts of traffic congregate. For example, if a network contains a segment made up of RISC-based platforms for engineering CAD, the files transferred to and from this segment by a server can be 2 MB or larger, and can therefore easily create gridlock on a 10-Mbps Ethernet network. Client/server database applications, particularly because they often have been implemented as replacements for mainframe-based implementations of such tasks, can generate heavy network traffic as well.

These conditions and more have helped give rise to a number of new, high-bandwidth networking technologies. Among them are:

- the PCI bus
- 100 Base T Fast Ethernet
- Asynchronous Transfer Mode, or ATM
- the copper-wire derivative of FDDI, called TP-PMD
- 100VG-AnyLAN
- switched Ethernet

We examine each of these technologies in this chapter..

The PCI Bus

Although it might seem out of place to begin a discussion of improving the performance of network hardware with a review of bus architecture, the topic is actually quite germane. The relatively new bus architecture called *peripheral component interconnect* or PCI offers several network-performance advantages over the older EISA bus.

The PCI bus has an overall throughput of about 132 Mbps, as opposed to about 33 Mbps for EISA. What's more, PCI buses improve overall I/O in ways other than raw speed. PCI can separate the peripheral bus into a master and subsidiary, local buses. By doing so, it also limits data traffic to and from individual portions and even components of the I/O system, thereby eliminating the contention for the bus that can take place with EISA. As a result, multiple PCI NICs, when installed in a server, can help that server improve its performance by providing it a larger aggregate throughput. Such a result is even more apparent in those adapters that conserve server CPU cycles and attempt to ensure low CPU utilization.

- - - - - - - - - - - - - - - - - -

NOTE: *Many PC and workstation vendors, including Apple, Compaq, Dell, Packard Bell, Hewlett-Packard, and IBM, have made PCI their bus-of-choice for future servers, for just such reasons as these.*

Fast Ethernet

100 Base T Fast Ethernet, that is, 100 Mbps Ethernet, has a number of advantages over its 10 Mbps predecessor, including, besides the obvious performance improvements it offers, a broad base of support from and sources among networking technology vendors, as well as the ability to provide a cost-effective migration path that allows the retention of much of a network's existing equipment and cabling.

100 Base T Fast Ethernet's General Characteristics

100 Base T Fast Ethernet evolved from 10 Base T Ethernet. Like 10 Base T, Fast Ethernet uses the *carrier sense, multiple access/collision detection* (CSMA/CD) media-access method, requiring, therefore, no protocol translation beyond that already needed in an environment. As a result, 100 Base T can be phased into a network. Indeed, it is possible to create heterogeneous networking configurations, in which 100 Base T and 10 Base T segments coexist quite peacefully, thanks to the addition of a simple bridge connecting the diverse segments.

As an example of the kinds of bandwidth bonuses 100 Base T can present, consider this scenario. Imagine an existing 10 Mbps network that, on the average, runs at about 60 percent overall utilization. Such a network could not be greatly expanded; a transmission profile of this sort leaves little bandwidth for additional nodes. However, if even some of the segments of this network were upgraded to 100 Base T, overall network load would drop dramatically, leaving significant additional bandwidth available for either existing or planned applications and stations. 100 Base T Fast Ethernet accomplishes such feats without mandating major changes in a network's architecture; it supports the same unshielded twisted-pair wiring and fiber-optic cabling already found in existing 10 Base T networks.

A less obvious but still notable benefit offered by 100 Base T Fast Ethernet is its ability to protect an organization's investment in Ethernet training, support, and knowledge, since 100 Base T's operating characteristics so closely resemble those of 10 Base T. In similar fashion, 100 Base T is attractive to network administrators because it is the most cost-effective high-speed networking technology. 100 Base T Fast Ethernet is appreciably less expensive than other high-speed transmission media.

100 Base T's Origins and Ongoing Development

In August 1993, an organization, now made up of 75 vendor members, called the *Fast Ethernet Alliance*, was formed. It was this group that developed and continues to enhance Fast Ethernet and its related technologies. At the time this book was being written, the Alliance included, among its membership:

- hub vendor Asanti Technologies
- NIC vendors SMC and 3Com
- semiconductor vendors National Semiconductor and Intel
- workstation manufacturer Sun Microsystems

In June 1995, the IEEE adopted, as the basis for a new set of standards for 100 Base T, the following four specifications, proposed by the Fast Ethernet Alliance:

1. *100 Base TX*, a specification for running 100 Base T Fast Ethernet at 100 Mbps over two pairs of data-grade wire, that is, over ISO 11801 Category 5 UTP and Type 1 shielded UTP.

2. *100 Base T4*, a specification for running 100 Base T Fast Ethernet at 100 Mbps over four pairs of Category 3, 4, or 5 UTP.

3. *100 Base FX*, a specification for running 100 Base T Fast Ethernet at 100 Mbps over a two-strand fiber-optic cable.

4. *Media independent interface* or MII, a specification for how an 802.3 media-access controller interfaces with any of these 100 Base T

media. MII gives 100 Base T capabilities similar to those provided by the *attachment unit interface* (AUI) connection for 10 Mbps Ethernet.

100 Base T Fast Ethernet in More Depth

The IEEE's 100 Base T Fast Ethernet specification combines 10 Mbps Ethernet's CSMA/CD method with a suite of 100 Mbps physical layers.

Fast Ethernet uses existing 802.3 MAC layer interfaces connected through an MII to the physical layer. What 100 Base T does differently is to increase the speed of the MAC layer to 100 Mbps. At the same time, 100 Base T leaves the rest of an Ethernet MAC layer's characteristics, including factors like:

- packet format
- packet length
- error control

unchanged and therefore identical to those used by 10 Mbps Ethernet, thereby further facilitating transitions to or the inclusion of Fast Ethernet in existing architectures.

As mentioned in the previous section, 100 Base T supports a wide variety of till-recently-only 10 Mbps Ethernet cabling, such as fiber, two- and four-pair UTP, and two-pair STP. What's more, the Fast Ethernet specification allows network planners to mix and match, by permitting mixing any combination of the supported media as long as they interact through an appropriate switch or repeater.

NOTE: *This ability to mix media is reminiscent of that provided by 10 Base T when it coexists with, for instance, 10 Base 2 or 10 Base 5.*

100 Base T integrates with any form of 10 Base T through a technique, defined in the IEEE standard, called *autonegotiation,* which allows an adapter or switch capable of data transfer at both 10 Mbps and 100 Mbps rates to use the fastest rate supported by the devices it connects. Autonegotiation offers two important benefits:

1. It connects at the highest, but still an appropriate, speed, without user intervention.

2. It will not attempt a connection if the devices it's attempting to converse with share no common transmission technology or rate, thereby minimizing delays caused by failed transmission attempts.

100 Base T Topologies

Although 100 Base T preserves 10 Base T's 100-meter maximum UTP cable runs from hub to desktop, its different handling of the MAC interface mandates several network topology characteristics, such as:

- a maximum of 205 meters for a station-hub/hub-station connection's cable run in a two-repeater network
- a maximum of 305 meters for a mixed UTP/FDDI cable run in a single-repeater network; this maximum assumes a class II repeater
- a maximum of 400 meters for a fiber cable run that connects one *data terminal equipment* or DTE device such as a switch or bridge to another; this figure assumes half duplex 100 Base FX
- a *collision zone,* that is, the area of a network within which collisions can be expected to occur, 205 meters in circumference around connecting devices

100 Base T's Cabling Schemes

100 Base T Fast Ethernet implements three media specifications:

1. 100 Base TX
2. 100 Base T4
3. 100 Base FX

Table 11-1 outlines the characteristics of these cabling specifications.

ATM

Although ATM offers transmission rates from 25 Mbps to more than 620 Mbps, this newest of the high-speed technologies, because it is quite expensive, is suitable only for backbones and not for desktop connectivi-

TABLE 11-1

Understanding
100 Base T Cabling

This Fast Ethernet Cabling	Uses This Wire	Operates in This Way
100 Base TX	Two-pair Category 5 UTP, which was designed for high-speed data transmission	One pair transmits, at about 125 Mhz; the other detects collisions. The physics of high-frequency signals like the 125 MHz signals put out by 100 Base TX is touchier at 100 Mbps than 10 Mbps. As a result, 100 Base TX requires more sophisticated encoding than 10 Base T.
100 Base T4	Four-pair Category 3, 4, 5 UTP. Note that Category 3 performs poorly at frequencies above 25 MHz, where high-speed data transmission takes place, and fails to meet FCC or European radio/EMI emission standards.	A data signal is split among the four wire pairs by a scheme that uses three pairs for transmission and the fourth for collision detection. 100 Base T4 permits 100-meter cable runs, but, unlike similar 10 Mbps cabling, can't employ 25-pair bundles. What's more, its asymmetrical nature makes it unsuitable for the full duplex configuring that allows 100 Base T to extend beyond the 205-meter limit of the current standard. Full duplex requires the ability to transmit 100 Mbps in both directions simultaneously. Since 100 Base T4 uses three of its four pairs to transmit in one direction, and only one pair to receive, this scheme isn't fully 100-Mbps-capable.
100BaseFX	Two-pair 62.5 to 125 micron fiber	Uses signaling similar to that of 100 Base TX. 100 Base FX's greatest advantage: the ability to transmit data over greater distances than UTP, making it particularly useful for backbone connections between bridges, routers, and switches. 100 Base FX uses MIC, ST, or SC fiber connectors like those defined for FDDI.

ty. It's the design that underlies ATM—the assumption that it will carry real-time interactive multimedia applications made up of any combination of voice, video, and data—that contributes to its cost, which can be as much as four times that of 100 Base TX.

But ATM has an even bigger drawback. It is still well short of standardization, and, as a result, ATM products from different vendors are likely to experience interoperability problems. In addition, integrating ATM into existing Ethernet environments may require ATM hardware that can emulate the networking subsystems of certain OSes, or the modification of those subsystems to accommodate ATM.

FDDI/TP-PMD

The *fiber distributed data interface* (FDDI), which delivers standards-based 100 Mbps performance over fiber-optic cable, is widely deployed in backbone installations. Its copper-based derivative, *TP-PMD,* uses a variation of the FDDI MAC to provide 100 Mbps over copper cabling.

TP-PMD, however, like ATM, is an expensive desktop solution, costing three to five times more than 100 Base T Fast Ethernet. What's more, it is difficult to implement and support. As a result, TP-PMD will, at least for the time being, remain primarily a backbone medium.

100VG-AnyLAN

Like 100 Base T Fast Ethernet, 100VG-AnyLAN offers 100 Mbps data throughput. Unlike Fast Ethernet, though, this technology has only one significant advocate, Hewlett-Packard.

100VG-AnyLAN relies on an access method called the *demand priority access method* or DPAM, which differs radically from Ethernet's CSMA/CD in using a round-robin, two-level priority scheme. Migrating to 100VG-AnyLAN would therefore force network support personnel to relearn their craft. This fact, taken together with its being offered by only one major vendor, give 100VG-AnyLAN only limited potential.

Switched Ethernet

An Ethernet switch, which connects segments of a larger overall net, can increase the aggregate bandwidth, but limits any individual user connection to a maximum of 10 Mbps. So, applications like real-time multimedia, which require more throughput than this, can't benefit from the use of an Ethernet switch. Ethernet switches can, however, be added to existing Ethernet networks without having to change anything other than this single component. Nevertheless, any node connected to a 10 Mbps Ethernet switch can never achieve throughput greater than 10 Mbps.

Cisco's Fast EtherChannel

For the balance of this chapter, we review one prominent commercial implementation of Fast Ethernet: *Fast EtherChannel* from Cisco Systems.

This technology builds upon standards-based 802.3 full-duplex Fast Ethernet to provide scalable bandwidth between 200 Mbps and 800 Mbps. Fast EtherChannel does so by:

- grouping multiple full-duplex point-to-point links, each of which is capable of full-duplex sensing of and negotiation with the other links in the group.

- being able to be placed at almost any point in a network, including between routers and switches, or between such devices and servers.

- being able to be configured flexibly within a single network. For example, Fast EtherChannel can be implemented at 400 Mbps between wiring closet and data center, but at up to 800 Mbps between servers' backbone.

- offering load-balancing traffic across the links it manages, thereby allowing the even distribution of unicast, broadcast, and multicast traffic.

- automatically providing redundant parallel transmission paths within a network, and as a result quickly and without user intervention redirecting traffic in the event of the failure of a specific link; Fast EtherChannel usually accomplishes such redirection in less than a second. As a result, protocol timers seldom expire, and sessions are not dropped.

- being transparent to network applications, and therefore requiring no modification of them.

Fast EtherChannel Components

Fast EtherChannel is actually a set of technologies, which includes:

- two to four industry-standard Fast Ethernet links that provide load sharing, up to 800 Mbps usable bandwidth, and that can be used with Fast Ethernet running on UTP, single-mode, or multimode fiber cabling.

- no need for the 802.1D *Spanning-Tree Protocol* (STP) to maintain topology state within the channel; since it uses a peer-to-peer control protocol that provides autoconfiguration and subsecond convergence times for parallel links, while at the same time permitting higher-level protocols such as STP, or existing routing protocols, to maintain topology. As a result, Fast EtherChannel can maintain or even improve upon a network's routing recovery capabilities without adding undue complexity to an environment or creating incompatibilities with third-party equipment or software.

- being easily configurable, either through the command line interface provided by Cisco or through SNMP applications like that offered by Cisco itself, illustrated in Fig. 11-1.

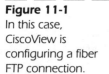

Figure 11-1
In this case, CiscoView is configuring a fiber FTP connection.

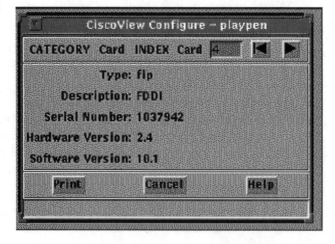

Fast EtherChannel Scenarios

Cisco Systems offers a number of examples of possible implementations of Fast EtherChannel, including the following.

Using Fast EtherChannel, a network can increase available bandwidth between wiring closets and data center from 200 to 400 Mbps and at the same time provide subsecond convergence in the event of a link's failing.

A network manager can increase bandwidth between data center and wiring closet to an aggregate of 800 Mbps and at the same time use the physical diversity of the fiber plant to decrease the chances of a network outage. This implementation of Fast EtherChannel relies on four fiber Fast Ethernet cable runs configured as two pairs, with each pair providing 400 Mbps of the overall bandwidth available. With such a configuration, should one member of one pair, or even one entire pair, be damaged, the remaining pair can pick up the damaged pair's traffic in less than a second, without any loss of user sessions.

A network has been completely designed around Fast EtherChannel. As in the previous examples, links from wiring closets into the data center run at 400 Mbps. Within the data center, routers are connected with Fast EtherChannel, thereby providing them with more bandwidth for inter-subnet routing. And since Fast EtherChannel is based on Fast Ethernet, there are no higher layer protocol incompatibilities within the router; performance improves regardless of protocol type. Finally, this network design also includes a server attached to the network by means of a four-link Fast EtherChannel, providing that server 800 Mbps of bandwidth to the network.

IN A NUTSHELL ▪▪▪ ▪▪ ▪▪ ▪▪ ▪ ▪▪

1. The PCI bus improves intrasystem communication by as much as 400 percent, moving data at about 132 Mbps, in contrast to EISA bus speeds of about 32 Mbps. As a result, PCI NICs can contribute to better throughput.

2. Any of the varieties of Fast Ethernet offer not only ten times the possible throughput, but also greater recoverability and enhanced routing abilities than do any varieties of 10 Mbps Ethernet.

3. Additional high-speed networking technologies such as ATM and

TP-PMD, because they are quite costly, are at this time suitable only for use in backbones.

LOOKING AHEAD

Chapter 12 begins with a brief theoretical discussion of simulating network loads as a means of gathering the information needed to improve network performance. Then the chapter applies those and other concepts in a specific context: improving the performance of Windows NT.

Tweaking an Operating System to Maximize Bandwidth

We begin this chapter with a very quick look at the kinds of data which underlie optimizing bandwidth use. Then we proceed to examining an example of doing just that, by improving the network-related performance of an operating system: Windows NT.

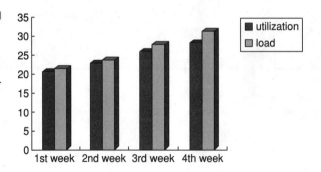

Figure 12-1
These are the indicators of a healthy network.

Network Load

Exactly what is the relationship between the theoretical load offered to a network by its various nodes, and real network utilization? Ideally, utilization should increase in a linear way with increases in load, as Fig. 12-1 indicates.

Transfer delay should also increase proportionally, not at a much larger rate than, load. If the latter is not the case, bandwidth, however much might remain available, is not being used to its potential. A number of factors, including transmission delay, collisions, and flexibility of routing, can affect real network utilization.

Windows NT

Among the organizations most involved in unraveling bandwidth tangles is Cisco Systems. In this section, we examine the results of a Cisco study on the efficient configuration of routing for NT-based nets. Then, later in the section, we turn to suggestions from Microsoft for fine-tuning the network performance of this OS.

Optimizing Network Performance by Optimizing Routing

IGRP, a routing protocol devised by Cisco Systems, is the keystone of Cisco's work with NT. Among IGRP's bandwidth-related characteristics are:

- low overhead; the protocol uses no more bandwidth than is actually needed to accomplish a task

- minimization or even elimination of routing loops

- simultaneous operation of any number of routing processes

NOTE: *Routing protocols, like other networking software, can be installed and configured through the Windows NT Control Panel.*

IGRP allows routers to maintain routing tables dynamically, through information gained by conversing with other routers. The protocol was designed to operate in networks with complex architectures and heterogeneous bandwidth and delay characteristics.

Autonomous Systems. IGRP uses, as one of its fundamental tools, the concept of an *autonomous system,* which is a collection of networks that share a common administration as well as a common routing strategy. By this definition, therefore, an autonomous system can be made up of one or many segments or even networks. Further, none of the components of an autonomous system need be divided internally, that is, use subnetting.

Every autonomous system has defined for it by a network administrator a unique identifying number; running IGRP requires using this number.

Interior, System, and Exterior Routes. As you might imagine, given the fact that IGRP builds and maintains its routing tables dynamically, it regularly advertises routes.

NOTE: *Recall from Chap. 6 that to a router, to advertise means to exchange information with its peers regarding routing conditions.*

IGRP advertises three types of routes:

1. *interior* or *subnet routes* of the network to which the IGRP update is being sent

2. *system routes,* that is, major routes *within* an autonomous system

3. *exterior routes,* designated so either by administrative command or through an update received from another router

IGRP, in handling any of the categories of routes it recognizes, acts as a distance vector protocol. Such protocols ask every router to send to each of its neighbors, at defined regular intervals, at least some and in many cases all of its routing table. It is this regularly exchanged routing information that routers use to calculate distances to all nodes. IGRP factors a variety of quantities into its measurements of these distances, including:

- internetwork delay
- raw bandwidth
- segment or network reliability, that is, physical condition
- load

Now, here's the really neat part. With IGRP, a network administrator can specify the weight each of these factors will have in the protocol's calculations of optimal routes.

IGRP or RIP?

Why use IGRP over a more traditional routing protocol such as RIP, in an NT or any network? The best reason, as well as the most basic difference between IGRP and the routing information protocol or RIP can be found in the makeup of the algorithms these protocols use in calculating optimum routes.

NOTE: *The combination of factors that make up such a calculation is known, in routing and in general networking parlance, as a metric.*

Table 12-1 compares these calculations. Most important about route calculations based on these factors and the ranges of values available to them is the flexibility provided to the network administrator in the configuration of heterogeneous networks, allowing him or her to:

- tailor the characteristics of individual segments to the nature and historical behavior of those segments
- combine these factors in various ways to fine-tune segment interaction and the network's overall performance

As Table 12-1 demonstrates, RIP, because it was designed for small, homogenous networks, cannot provide the flexibility offered by IGRP.

TABLE 12-1

IGRP, RIP, and the
Calculation of
Routes

The Measurement	Represents	Used by RIP?	Used by IGRP?	In the Range
Bandwidth	Throughput in bits per second of which the slowest link in a path is capable	No	Yes	1200 bps to 10 Gbps
Hop count	Number of routers and/or paths a message traverses before reaching its destination	Yes	No	1–15
Path channel occupancy	How much of the bandwidth available to a path is actually in use	No	Yes	Integer values in the range 1–255
Path reliability	Error rate, that is, the ratio of packets reaching their destination undamaged compared to the expected number of such packets; the higher the ratio, the more reliable the path	No	Yes	Integer values in the range 1–255
Topological delay	The amount of time needed to reach a destination along a specified path	No	Yes	1 to 224 milli-seconds

Optimizing Network Performance by Fine-Tuning NT

A drop-off in network performance can result from a network's simply being overloaded or being physically damaged. However, Microsoft agrees with the researchers at Rice University, whose work we reviewed in Chap. 7, that server configuration, and particularly the distribution of critical server resources such as memory, can also make or break throughput.

Under Windows NT Server (3.51 and higher), one of a server's most important responsibilities is to establish sessions with remote stations, and to accept and fulfill *server message block* or SMB requests, as defined in Fig. 12-2, from them.

Figure 12-2
Under NT, network services take place one SM block at a time.

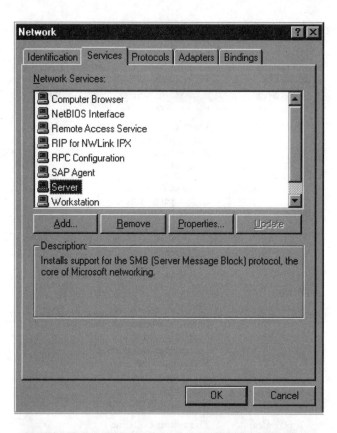

Figure 12-3
The Redirector, which we discuss in more detail later in this chapter, helps clients exchange service requests with NT Server.

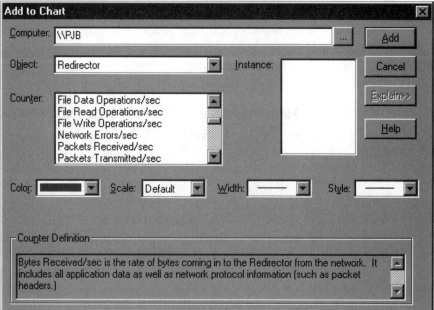

Such requests most frequently ask the server for I/O operations, like those shown in Fig. 12-3, involving a device or file located on the server.

Clearly, the efficiency with which such requests are carried out can be significantly affected by the way in which server resources, particularly memory, have been configured.

Windows NT Server resource allocation and the associated use of nonpaged memory, whose difference from paged memory Fig. 12-4 clarifies, can be configured through the NT Control Panel tool called *Network*.

Through this tool, you can specify the amount of memory that will be available to such server tasks as managing as work items, threads, and connections. Depending on the choices you make in the dialog shown in Fig. 12-5, the efficiency of such services can differ dramatically. Table 12-2 outlines what you can do with this dialog.

Now let's examine, through Table 12-3:

- some of what you might expect to see in the NT Server Event Log if the memory parameters we just examined aren't well configured

- what you can do about it

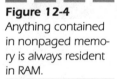

Figure 12-4
Anything contained in nonpaged memory is always resident in RAM.

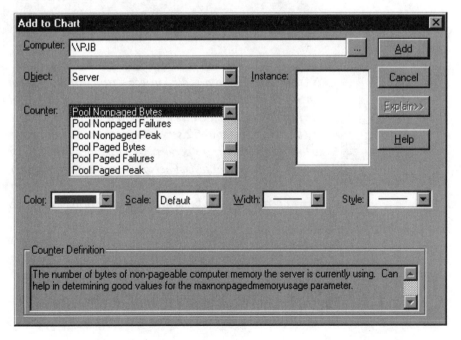

Figure 12-5
This dialog is the key to efficient use by an NT Server of its memory.

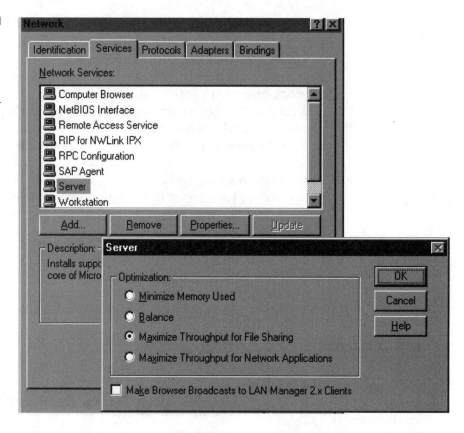

NOTE: Analogous paged/nonpaged settings are available under NT Server for other file system services, such as that which distributes Macintosh files. There, Mac-FilePagedMemLimit defines the maximum amount of paged, that is, virtual, memory that the Macintosh file server can use. As you can imagine given the discussion to this point, that server's performance improves with an increase in this value. Increasing the value of MacFileNonPagedMemLimit, which specifies the maximum amount of real RAM available to the Macintosh file server, also improves that server's performance but may injure the performance of other system resources.

If Performance Monitor indicates that processes other than those generated by the Server, such as the Spooler, are competing with the server for processor time, you have two workarounds available:

1. increasing the priority at which server threads run
2. decreasing the priority of the non-Server CPU glommer

TABLE 12-2

Windows NT Server Memory Management

This NT Server Parameter	Helps Maximize Server Efficiency in Handling
Minimize Memory Used	Up to 10 simultaneous remote user connections, as Fig. 12-6 shows.
Balance	11 to 63 simultaneous remote user connections, as Fig. 12-7 indicates.
Maximize Throughput for File Sharing	64 or more remote connections, *and* at the same time giving file cache access to memory priority over user application access to memory, as Fig. 12-8 illustrates. Changes the value of the Registry parameter LargeSystemCache to 0×1. This option, the default, optimizes NT Server's file server performance.
Maximize Throughput for Network Applications	64 or more remote connections, *and* at the same time giving user application access to memory priority over file cache access to memory, as Fig. 12-9 indicates. Changes the value of the Registry parameter LargeSystemCache to 0×0. Such a setting is of most use to applications that do not carry out their own memory management. An example of one that does? Microsoft's SQL Server.

Or, since every remote connection made by a client to NT Server consumes at least a small amount of memory, you could improve the overall effect of NT on networking and throughput by fine-tuning the parameter Autodisconnect. By reducing the value assigned to this parameter, which defines the time interval after which inactive connections that involve no open files are terminated, you can at least somewhat facilitate the server's ability to handle active sessions.

NOTE: *Many of the performance counters available to Windows NT Server were introduced in version 3.51; these include active threads; available threads; available work items; bytes received, transferred, or sent per second; current clients; read bytes and operations per second; and write bytes and operations per second. Taken together with the performance counters already present in that release, such as blocking requests rejected; bytes received, transmitted, or total per second; errors logon; files open; logon total; and many more, these counters make the fine-tuning of any recent release of Windows NT Server possible.*

Figure 12-6
Here's what's involved in minimizing NT Server's use of memory.

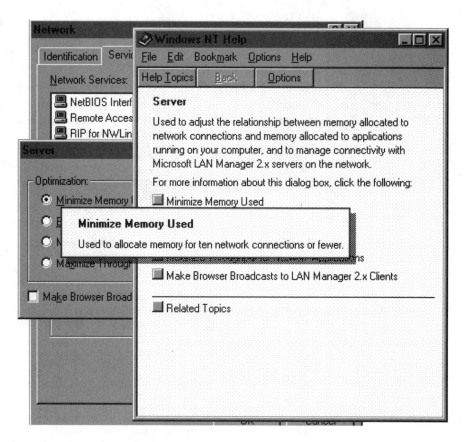

The Redirector. When Connect, Open, Read, or Write requests involving files whose pathnames indicate a redirected, that is, network, drive such as one defined by the command

```
net use x: \\my_srvr\my_shrd_drv
```

those requests are forwarded to the Windows NT redirector, which in turn hands off the request to the appropriate transport layer protocol, be it TCP/IP, NBF, or NWLINK, which actually places the request onto the transmission medium. From there, the server retrieves it and responds. While this sequence of events might seem to reaffirm the central role of the server in overall network performance, it also indicates that the redirector side of the conversation has an effect on that performance, too. Table 12-4 outlines some Redirector-related performance counters, and how they can be part of optimizing the Redirector's effect on throughput.

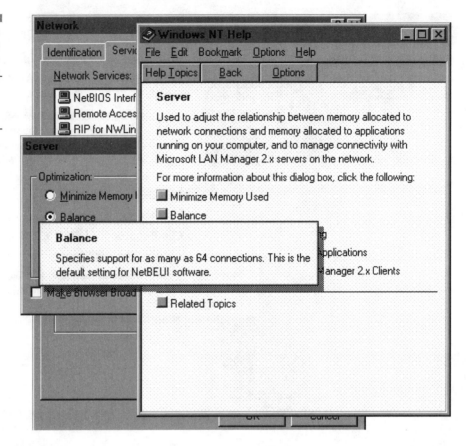

Figure 12-7
Balancing NT Server's use of memory is important when dealing with multiple simultaneous client connections.

Netlogon. Netlogon is the NT service that keeps all user account databases synchronized, that is, the same, whether they reside on an NT primary or backup domain controller.

■■ ■■ ■■ ■■ ■■ ■■ ■■ ■■ ■■ ■■ ■■ ■■ ■■ ■■ ■■ ■■ ■■ ■■ ■■

NOTE: If you're unfamiliar with NT's concept of primary versus backup domain controllers, here it is in a nutshell. A primary controller, of which a given network can have only one, is:

- *by default the first server installed in a new NT network*
- *the repository for all user, group, and machine information*

A backup controller, however many there might be in a domain:

- *contains a copy of this database, updated periodically*
- *can carry out tasks like authenticating users*

▬▬ ▬▬ ▬▬ ▬▬ ▬

Figure 12-8
The assumption here is that a large network implies a high level of file sharing.

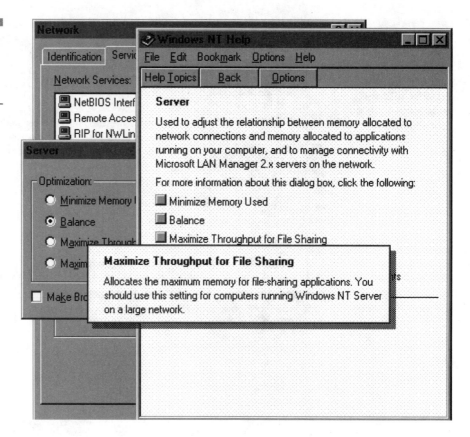

If, in eyeballing Performance Monitor's output, you should discover that your Primary Domain Controller (PDC—love those Microsoft acronyms!) is experiencing inordinate amounts of maintenance-related traffic, you can compensate to decrease those traffic levels. On your PDC, increase the interval represented by the Netlogon-related parameters:

- service update notice period, or the length of time the PDC will wait before updating system information on backup controllers or BDCs

- server announcement period, or the length of time the PDC needs to make BDCs aware it's there

Figure 12-9
It's important to make
this choice if your NT
Server distributes
applications which
do not do their own
memory manage-
ment.

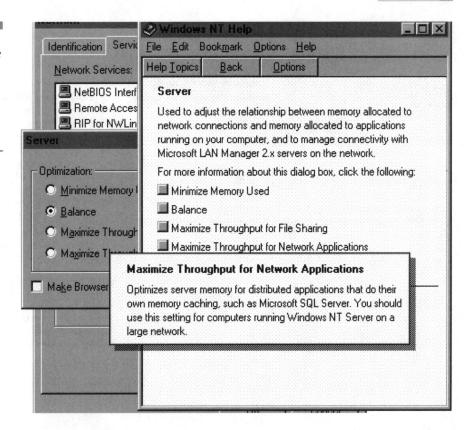

To do so, you must use the registry editor REGEDIT.EXE to change
the path

`HKEY_LOCAL_MACHINE\SYSTEM\CurrentControlSet\Services\Netlogon\Parameters`

as indicated in Table 12-5.

*NOTE: In addition to the points covered in Table 12-5, keep in mind the
amount of traffic generated by even a single change to the user account database.
A change of password or the addition of a new user creates about 1 KB of data.
Any changes involving user groups produces about 4 KB of data. If such transactions take place frequently, they too can place an added strain on your NT-based
network. But take heart. Netlogon's actions are also affected by a control called*
`REPLICATIONGOVERNER`, *which prevents Netlogon from using more than 25
percent of the total bandwidth available to the server.*

TABLE 12-3

Using the Event Log to Trouble-shoot NT Server Memory Use

This Error	Which Indicates	Can Be Corrected or Improved Upon by
2009	Insufficient server storage available to process a request to expand a table.	Using Performance monitor to track the counters Server/WorkItemShortages (WorkItem = location used to store an SMB), Server/Pool Paged/Nonpaged Failures, and Server/ContextBlockQueueTime. If the first two of these increase steadily, or if the last regularly averages more than 50 milliseconds, there is a bottleneck at the server in handling remote I/O requests. Whether such gridlock is caused by a poorly chosen optimization level for NT Server, or by glutted disk, CPU, or memory, your best bet may be simply to locate and off-load or otherwise redistribute the resource-gluttonous tasks.

Or, if Server/Pool Paged/Nonpaged Failures increase, indicating that the server is running out of memory as it was originally allocated, you can increase memory available by using either or both MaxNonPagedMemoryUsage and MaxPagedMemoryUsage. |
2017	NT was unable to allocate from the nonpaged pool because the configured limit for allocation from that pool had been reached.	
2018	NT was unable to allocate from the paged pool because the configured limit for allocation from that pool had been reached.	In all these cases, the most likely explanation is that your server lacks adequate physical memory. Look at it this way—RAM is cheap enough.
2019	NT could not allocate from the nonpaged pool because that pool was empty.	
2020	NT could not allocate from the paged pool because that pool was empty.	

TABLE 12-4

Improving Redirector's Performance

This Counter or Message	Indicates	Which in Turn Indicates
Redirector Current Commands	The number of requests to the Redirector currently queued for service	*If* this number is much larger than the number of network adapter cards installed in the server, that the network and/or the server are seriously traffic-jammed. Rather than installing more NICs in an effort to clear up this problem, try increasing the maximum number of pending network commands allowed.
Redirector Network Errors/sec	SMB requests are timing out, forcing the redirector to disconnect, reconnect and recover	You should try to increase the value of the SessTimeout Registry parameter, which defines the maximum amount of time that the redirector allows an operation that is not long-term to remain outstanding, from its default of 45 seconds.
ERROR_INVALID_USER_BUFFER	Too many outstanding asynchronous I/O requests	You should try to increase the redirector's thread count
ERROR_NOT_ENOUGH_MEMORY	Too many outstanding asynchronous I/O requests	You should try to increase the redirector's thread count.
Provider Order	As displayed in either the Registry entries: SYSTEM\CurrentControlSet\Control\NetworkProvider\Order:ProviderOrder or the Control Panel's Network tool, the order of DLLS in which a network-related API call will be routed. Applicable only if more than one Redirector has been installed and configured, as, for example, on a server which, in addition to offering NT client services, provides Client Services for NetWare as well.	You might want to monitor the individual redirector's thread counts, and adjust them if it appears that will improve overall server performance.

TABLE 12-5

Editing Netlogon Registry Parameters to Improve Performance

The Parameter	Represents	Has the Default	A Minimum Allowed of	A Maximum Allowed of	And the Effect on Traffic
Pulse	An interval during which all user account- and security-related database changes made since the last transmitted signal to BDCs are collected, and at the completion of which a pulse, that is, a notifying signal, is sent to BDCs which need to be updated. BDCs respond by asking for any database changes.	300 seconds	60 seconds	3600 seconds	Increases here can lessen traffic during the interval, but increase it when the larger resulting set of changes must be transmitted. Decreases might help stabilize traffic patterns.
PulseConcurrency	The maximum number of simultaneous pulses the PDC will send to BDCs	20	1	500	Increasing PulseConcurrency increases the load on the PDC, and therefore demands for such resources as memory. Decreasing PulseConcurrency increases the time it takes for all BDCs to have their user account databases updated, and therefore the possible levels of related network traffic.
Randomize	The BDC back off period, that is, the number of seconds, between 0 and the possible maximum value of 120, for which the BDC will wait, after receiving a pulse, before calling its PDC.	1 second	0 seconds	120 seconds	Either increasing or decreasing Randomize without an eye to the patterns of calls by BDCs that will result might actually contribute to bottlenecks, since, as a result, more rather than fewer BDCs might try to get the PDC's attention at the same time. Balance this parameter across all BDCs.

Print Browser. When a printer is shared through Windows NT, the printer spooler creates a thread that broadcasts a message, with all of the protocols configured for the print server, to all other NT print servers, making them aware of the newly shared printer. Each of the print servers so notified adds to its local printer-browse list the name of the newly shared printer. Then, each notified print server rebroadcasts the list of its local printers to all other print servers every 10 minutes. All of these broadcasts, although a laudable effort to ensure that all NT print servers have up-to-date browse lists, clearly generate significant network traffic. If you don't mind print servers being oblivious of their peers across the network, you can eliminate at least this variety of traffic by editing the Registry to disable the printer browse thread. Update the path

```
HKEY_LOCAL_MACHINE\SYSTEM\
CurrentControlSet\Control\Print\Providers
ServerThreadRunning (default = 0 REG_SZ)
```

to a value of 1 rather than the default of 0.

Segmenting. Should the counter Server Bytes Total/sec, representing the number of bytes sent to and received from the network by an NT server, approach or equal a network's maximum theoretical transfer rate, it may be time to segment that network. For instance, if the counter in question shows a value of more than 1,100,000 (remember, we're talking byte now, not bits—divide 10,000,000 by 8) on a 10-Mbps Ethernet network, it's probably time for a reconfiguration of this sort.

If you must segment, keep the following rules of thumb in mind:

1. Match any new adapter to the server system bus; if that is 32 bit, for example, use a 32-bit network adapter.

2. Avoid sequences of fast-to-slow adapters.

3. If you haven't yet done so, install and use the Network Monitor. Doing so provides you with additional Performance Monitor counters, as well as statistics generated by the tool itself, such as:

   - *% Network Utilization,* the percentage of total available network bandwidth being used

   - *Frames per Second,* the number of frames being transmitted across the network per second

   - *Bytes per Second,* the number of bytes being transmitted over the network per second

   - *Broadcasts per Second,* the number of broadcast frames pushed onto the network per second

- *Multicasts per Second,* the number of multicast frames sent over the network per second
- *Network Card (MAC) Statistics,* the cumulative total number of frames, bytes, broadcasts, and multicasts seen on the network by a given network card since it began functioning
- *Network Card (MAC) Error Statistics,* the cumulative errors, including CRC Errors and frames dropped either because no buffer space was available or because of hardware problems, seen by a given network card since it began functioning

In particular, increases in Broadcasts or Multicasts per second can affect both server and node performance. Each such transmission causes every NIC on the network to generate an interrupt that will cause broadcast or multicast messages to be passed up to the appropriate transport layer protocol. Such interrupts and their processing can hobble a server's CPU. Broadcast or multicast rates of more than 100 per second should cause you to consider such actions as configuring a router to disable TCP ports 137 and 138, the destinations for these types of traffic. However, even this step may not completely solve the problem, since one of NT's available transport protocols, NBF, is not in any case routable.

The counter Network Segment % Network Utilization, which represents that portion of total network bandwidth being consumed by a given segment, is another parameter to monitor closely. Should this value reach or exceed 40 percent, the network is once again gridlocked, and may have to be either further subdivided, or redivided according to a different pattern.

Since, particularly in a growing or large network, the likelihood that two hosts will try to transmit simultaneously increases as overall network use increases, collisions can be a very effective reflection of network load. To use collisions in this way, look for a total that doesn't exceed 15 percent of the total number of output packets. Should collisions exceed this figure, probably the only solution is to change the pattern of network segmentation so as to reduce traffic.

NOTE: *Ethernet networks begin to experience significant and handicapping rates of collisions at about 67 percent utilization, which on a garden-variety, 10-Mbps network translates to a throughput of about 6,700,000 bits or nearly 84,000 bytes per second.*

IN A NUTSHELL

1. Ideally, network utilization and network load should, as much as possible, have and increase by a one-to-one relationship.

2. In heterogeneous networks, the use of a flexible routing protocol like IGRP can help to optimize throughput.

3. As is the case with UNIX, the configuration of memory in a server running under Windows NT can have a significant effect on throughput.

4. In addition to fine-tuning server memory management, optimizing the configuration of such NT tools as the Redirector and Spooler can help to improve network performance.

5. The presence and pattern of segmentation used in a network can improve or weaken network performance.

LOOKING AHEAD

In Chap. 13, we investigate not how to improve an operating system's bandwidth management, but how to refine the relationship between bandwidth and server applications. Specifically, we look into ways to hone the performance of an HTTP implementation.

CHAPTER 13

Tweaking Network Applications to Maximize Bandwidth

We begin this chapter by examining the effect on demands for bandwidth of the habits and practices of its single largest category of consumers: end users. Then we turn to investigating recent research by a number of members of the World Wide Web Consortium (W3C) into how to satisfy those demands and at the same time improve network throughput.

Human Factors and Network Performance

Research on a wide variety of hypertext systems has shown that users need response times of less than one second when moving from one page to another if they are to navigate freely through an information space. Traditional human factors research into response times also shows the need for response times faster than a second. For example, studies done at IBM in the 1970s and 1980s found that mainframe users were more productive when the time between hitting a function key and getting the requested screen was less than a second.

Currently, the minimum goal for response times should therefore be to get pages to users in no more than ten seconds, since that's the limit of people's ability to keep their attention focused while waiting. Figure 13-1 shows the distribution of the speeds with which users have connected to the Internet, according to the last six surveys of Web users conducted by the Graphics, Visualization, and Usability Center at Georgia Tech.

Figure 13-1
Users often connect to the Internet at speeds less than those of which their modems are capable.

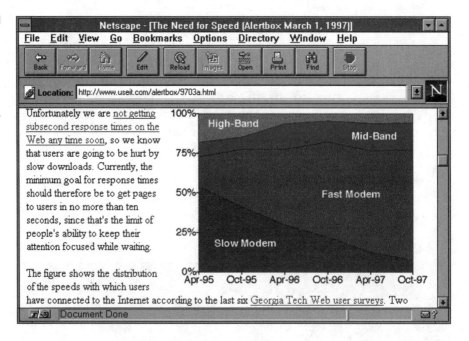

This figure makes three things clear:

- users are replacing slow modems with faster ones, but the proportion of users who connect at modem speeds stays about the same
- the proportion of users connecting at high bandwidths (T1 or better) is going down, even though the Web requires at least T1 speed to work well
- though use of midband connections such as ISDN has increased, such speeds are insufficient for decent Web response

Several links in the chain from server to browser affect throughput, and therefore the response times experienced by end users:

- server throughput, which can be degraded by rapidly increasing traffic
- the nature of the server's connection to the Internet (e.g., T1? T3?)
- bottlenecks on the Internet itself
- the manner in which the user connects to the Internet
- the rendering speed of browsers and client computers

The delays generated by these and other factors are cumulative. Therefore, improving the effect on throughput of one or even a few of these stages of the distribution of information may not significantly improve overall throughput and response time.

Given these characteristics of both human and network information processing, it becomes clear that Web pages should be designed with speed in mind, and should even consider speed of download the overriding design criterion. Here are some steps that can be taken to keep HTML document sizes small:

- Keep graphics to a minimum.
- Use multimedia effects only when they truly add to the user's understanding of the information being presented.
- Use style sheets to improve the effectiveness of pages without a corollary increase in download time. As much as possible, use linked rather than embedded style sheets, since the former is downloaded only once.
- Improve the speed at which a page initially loads by:

including WIDTH and HEIGHT attributes on all images and table columns, so that browsers have all the information they need to draw the top of pages as quickly as possible

since complex tables often take a long time to render, minimizing tables' complexity

when embedding URLs in Web pages, explicitly coding the final slash, in order to avoid server redirects and thereby reduce latency

HTTP/1.1, CSS1, PNG, and Network Performance

This section summarizes recent research by members of the W3C into improving throughput by:

- using version 1.1 of the HTTP protocol
- using style sheets to streamline Web publishing
- using compression, particularly PNG-based compression, to reduce the size of image files

The Nature of Web Pages

Typical Web pages today contain not only an HTML document, but also as many as 20 or more embedded images. Each of these images is an independent object in the Web, retrieved or validated for change separately. Most frequently, therefore, a Web client will:

- fetch the base HTML document
- immediately begin to retrieve any embedded objects

HTTP was not designed to handle large numbers of such objects. As a result, HTTP/1.0 handles multiple requests to the same server (such as those generated by the successive client retrievals cited previously) inefficiently, creating a separate TCP connection for each object being downloaded. The recently released HTTP/1.1 standard was designed to address this problem by encouraging multiple transfers of objects over one connection. Coincidentally, expected changes in Web content are expected to decrease the number of embedded objects, which, together with HTTP/1.1, will further improve network performance.

To test the effects of some of the new features of HTTP/1.1, members of the W3C simulated two varieties of client behavior:

- visiting a site for the first time, with an empty client cache
- revalidating cached items when a site is revisited

Tests were conducted in three different network environments:

- a high-bandwidth, low-latency LAN
- a high-bandwidth, high-latency WAN
- a low-bandwidth, high-latency dial-up PPP connection

HTTP/1.1 and HTTP/1.0

Among the most important differences between these two versions of the Web protocol is the way in which each interacts with TCP/IP. HTTP/1.0 opens and closes a new TCP connection for every operation it handles. This means that a significant percentage of the packets that make up server/browser traffic are TCP control packets that open or close connections. What's more, when a TCP connection is first opened, the protocol uses an algorithm known as *slow start*, which in turn causes the first several packets transmitted to examine the network in order to determine an optimal transmission rate. The use of the slow-start algorithm can therefore result in a Web object's being transferred before its underlying TCP connection has finished the associated slow start. Or, as the researchers from the W3C phrase it, *most HTTP/1.0 operations use TCP at its least efficient.* Much network congestion and unnecessary overhead ensue as a result.

HTTP/1.1 leaves the TCP connection open between consecutive operations. This technique, known as *persistent connection:*

- avoids the bandwidth costs of multiple opens and closes
- reduces the impact of slow start on bandwidth consumption

NOTE: *The Keep-Alive extension to HTTP/1.0 is a form of persistent connection. However, the design of HTTP/1.1's differs, albeit only in minor details, from Keep-Alive, primarily in dealing with problems that can result when Keep-Alive is used with more than one proxy between a client and a server.*

Persistent connections allow multiple requests to be sent without waiting for a response, which in turn can help improve throughput by minimizing round-trip delays, and thereby reducing the number of packets transmitted. This technique, an important factor in the design of HTTP/1.1, is called *pipelining*.

HTTP/1.1 also enables transport, as opposed to server- or client-application compression of certain data types, in order to enable clients to retrieve HTML or other documents, which are traditionally not compressed, in a compact form.

HTTP/1.1 Design Goals

The major design goals for HTTP/1.1 included:

- lowering the load on the Internet generated by HTTP, while at the same time reducing the congestion the protocol causes

- improving upon HTTP/1.0's caching, termed *primitive and error prone* by the W3C researchers, in order to enable applications to work reliably with caching

- improving the performance experienced by end users

NOTE: *HTTP/1.1 does not attempt to solve other common Web-related network performance problems, such as the intermittent transient network overloads experienced by high-use Web sites. But its designers feel that even scenarios like these can benefit from the new version of the protocol.*

Range Requests and Validation. To improve response time as it is perceived by users, a browser needs to learn basic size information of each object in a page that is needed for proper display of that page, as soon as possible. It's for this reason that the first bytes of an HTML document typically contain the image size.

In an attempt to retrieve the first few bytes of embedded links while still receiving such header bytes for a master document, HTTP/1.0 browsers frequently establish multiple TCP connections. HTTP/1.1, on the other hand, provides a *multiple range request* capability, which permits client browsers to achieve similar or better results over a single connection.

NOTE: *Although multiple range requests as they are to be implemented in HTTP/1.1 (and as they are already implemented in many current HTTP/1.0 servers) allow a client to perform partial retrieval of objects, the initial motivation for providing for such requests was to allow proxies to finish interrupted transfers by requesting only that part of a document not already in their cache.*

To provide browsers with adequate information on the size of embedded objects, the researchers at the W3C envision the revalidation request feature of HTTP/1.1 as combining cache validation headers and a range request header, in order to prevent large objects from monopolizing the HTTP connection to the server. Further, they anticipate the need for requesting a range large enough to return embedded metadata for objects of the common data types. Both these features have been included in the caching and range request capabilities designed into HTTP/1.1.

Changes to Web Content. In the scientific environment into which the Web was born, users were more concerned with content than with presentation. However, when nonscientific communities discovered the Web, the perceived limitations of HTML became a source of frustration for them. In the absence of style sheets, which have only recently begun to be accepted and used, authors tended, in the words of one W3C paper, "to meet design challenges by twisting HTML out of shape, for instance, by studding their pages with small images that do little more than display text."

As a result, the effects on throughput of such document design, and the effects on that design of continued growth in the use of style sheets—specifically, CSS1 style sheets—were also examined by members of the W3C.

Image formatting is another facet of Web page design investigated by W3C members. Most images distributed by the Web are in GIF format. Recently, a new image format, *PNG*, was introduced that has several advantages over GIF that are particularly relevant to bandwidth. These include:

- PNG images render more quickly on the screen.
- PNG can produce cross-platform images.
- PNG images are usually smaller than GIF images.

The characteristics and behavior of both PNG and its related animation format MNG, the latter more compact than animated GIF, were studied.

Tests of the Design of HTTP/1.1

As the basis for their tests of the effectiveness of HTTP/1.1, W3C researchers simulated an actively serving Web site. They supplied it data, made up of both HTML documents and GIF images, drawn from what they termed, in a gem of understatement, *two very heavily used home pages—* those of Netscape and Microsoft—and thereafter referred to as *Microscape.*

Microscape's Web Content.   Initially, the Microscape site contained only:

- a single page of typical HTML markup, weighing in at 42 KB
- 42 GIF images inlined in this page, which together totaled another 125 KB, and which, individually, ranged in size from 70 bytes to 40 KB, with 19 of the 42 images, one example of which Fig. 13-2 illustrates, taking up less than 1 KB, 7 using between 1 and 2 KB, and 6 needing between 2 and 3 KB

NOTE: *Such a page design, which results in a larger-than-needed HTML document, is nonetheless still commonplace on the Web today.*

Specific tests of HTTP/1.1 performed against this content were:

- *first-time retrieval,* simulating a browser visiting a site for the first time, that is, with its cache empty and having to retrieve the site's top-level page and all its embedded objects; for the Microscape Web site, equivalent to issuing 43 HTTP GET requests
- *revalidation,* simulating revisiting a home page whose contents are already available in the local cache, that is, simply validating that page and its embedded objects, rather than actually transferring them; for the Microscape site, equivalent to issuing 43 HTTP Conditional GET requests

solutions

Figure 13-2
Even this simple a GIF, weighing in at only 682 bytes, adds to demands for bandwidth.

Network Environments. W3C researchers used three combinations of bandwidth and latency to simulate the performance of HTTP/1.1 in commonly used network environments. These combinations are:

- a high-bandwidth, low-latency, 10-Mbps Ethernet LAN
- a high-bandwidth, high-latency WAN
- a low-bandwidth, high-latency dial-up PPP connection

HTTP/1.1 performance was tested in all the environments just noted under both:

- Jigsaw, a Web server written entirely in Java and relying, for control of TCP, on specific features of that language provided only by the Java Development Kit (JDK) 1.1
- Apache

Platforms used to test HTTP/1.1 included:

- server: a Sun Microsystems SPARC Ultra-1 running Solaris 2.5
- LAN client: a Digital Equipment Corporation AlphaStation 400 4/233 running UNIX 4.0a
- WAN client: a DEC AlphaStation 3000 running UNIX 4.0
- PPP client: a Pentium Pro PC under Windows NT Server 4.0
- HTTP server: Jigsaw 1.06 and Apache 1.2b10
- HTTP client: Netscape Communicator 4.0 beta 5 and Microsoft Internet Explorer 4.0 beta 1, both under Windows NT

Initial Tests. The first tests of HTTP/1.1 used:

- a robot set to use HTTP/1.0 requests, to assign one TCP connection to each request, and to permit a maximum of four simultaneous connections (the last chosen because it is Netscape Navigator's default)
- after the tests of HTTP/1.0 to which those of HTTP/1.1 would be compared, a robot acting as a simple HTTP/1.1 client using persistent connections, that is, with all communication taking place on the same single TCP connection

As Table 13-1 points out, these early tests of HTTP/1.1 against 1.0 showed 1.1 to use far fewer packets, but much more transmission time.

TABLE 13-1

A Jigsaw-Based,
High-Bandwidth,
Low-Latency Cache
Revalidation Test

	HTTP/1.0	HTTP/1.1 Persistent	HTTP/1.1 Pipeline
Max simultaneous sockets	6	1	1
Total number of sockets used	40	1	1
Packets from client to server	226	70	25
Packets from server to client	271	153	58
Total number of packets	497	223	83
Total elapsed time in seconds	1.85	4.13	3.02

Pipelining. In an effort to lower elapsed transmission time and improve its efficiency, the W3C researchers next introduced pipelining into *libwww,* the version of the standard library of Web-related system functions upon which HTTP/1.1 relies.

Without pipelining, an HTTP client will wait for a response to arrive before issuing new requests. With pipelining, as many requests as possible are issued at once. Under HTTP/1.1 with pipelining, responses are still serialized, and the structure of HTTP messages is the same. Only protocol timing is altered.

What's more, the altered libwww is very careful not to generate unnecessary headers and not to waste bytes on white space. The result is an average request size of around 190 bytes, which is significantly smaller than many existing product HTTP implementations.

Pipelined requests are buffered before transmission so that multiple HTTP requests can be sent with the same TCP segment. This practice has a significant impact on the number of packets required to transmit content; it also lowers demands on CPU time by both client and server. However, it's also inherent in pipelining that requests are not immediately transmitted. As a result, the W3C researchers devised methods to flush a Web server's output buffer.

At first, a version of libwww that used two such mechanisms was implemented. It:

- flushed the output buffer if the data there reached a certain amount. The W3C researchers settled on an output buffer size of 1024 bytes, since this was found to produce two full TCP segments under *maximum transfer units* or (MTUs) of either 536 or

512, or a single such segment with the common Ethernet MTU of 1460

- introduced a timer to the output buffer stream, by timing out after a specified period of time, which could force the buffer to be flushed

Interpreting and Expanding Upon First Results. After studying the mixed results of these first tests of HTTP/1.1, the W3C members involved in testing its design realized that those results were more influenced by the robot application than by libwww, since the robot has access to much more information about requests than the library. Therefore, the researchers introduced an explicit flush mechanism into the robot application, in an effort to significantly improve the performance of HTTP/1.1.

The robot was modified to:

- force a flush after the client had issued a single request for the test HTML document
- buffer the related requests for inlined images

NOTE: *HTTP libraries can be configured to flush buffers automatically after a timeout. But taking advantage of information available to a client application can produce a considerably faster protocol implementation than any that relies on such timeouts.*

Connection Management.   Because they permit only one connection at a time, HTTP/1.0 implementations can close both halves of a TCP connection simultaneously when they finish processing a request. A pipelined HTTP/1.1 implementation might cause serious problems if it did so.

Imagine this scenario. An HTTP/1.1 client talking to an HTTP/1.1 server pipelines a dozen or more requests on an open TCP connection. Should the server be set up to serve no more than five requests per connection, it would close the pipelined TCP connection in both directions after it successfully served the first five requests. The remaining 10 requests, already sent from the client, as well as associated TCP ACK packets, therefore will arrive at a closed server port. Such unanticipated data causes the server TCP stack to reset; this in turn forces the client TCP stack to pass the most recently acknowledged packet to the client application, *and then to discard all other packets.* In other words, HTTP responses that are either being received or already have been received but haven't been acknowledged will be dropped by the client stack.

In order to avoid such an obviously wasteful situation, the W3C researchers realized, a server must close each half of a TCP connection independently.

Web Content Representation

After making adjustments including those just outlined to HTTP/1.1 so that it outperformed HTTP/1.0 in transmission time as well as in packets used, the W3C researchers explored another avenue they felt would lead to better protocol and therefore overall throughput performance. Specifically, they examined using data compression of the HTTP message body.

This technique did not seek to compress HTTP headers. Nor did it compress images, since those were already compressed by algorithms such as GIF. Rather, it:

- compacted only the body of the message
- used the header *Content-Encoding* to describe the encoding mechanism

In doing so, this new technique relied on the *zlib* compression library version 1.04, freely available C-based code that provides a stream-based interface that interacts well with the libwww stream model.

NOTE: *PNG also uses zlib.*

HTTP/1.1 clients indicate that they can handle such content coding by sending a request that includes the header *Accept-Encoding: deflate.* In the W3C tests, a server carried out no on-the-fly compression. Instead, it sent a precomputed, deflated version of the Microscape HTML page. The recipient, on the other hand, did operate on-the-fly to inflate or decompress the document, which was then parsed with the client's usual HTML parser.

While the zlib library has several flags that allow the optimization of the compression algorithm, the W3C researchers used none of them, preferring to employ only default values for both deflating and inflat-

Figure 13-3

Whether handled by Jigsaw or Apache, the Microscape test HTML file, when compressed, resulted in reductions in number of packets and transmission time.

	Jigsaw		Apache	
	Pa	Sec	Pa	Sec
Uncompressed HTML	67	12.21	67	12.13
Compressed HTML	21.0	4.35	4.35	4.43
Saved using compression	68.7%	64.4%	68.7%	64.5%

Figure 13-4

In this table taken from the documentation of tests on HTTP/1.1, *Pa* represents packets sent, *Sec* is elapsed time in seconds to send them, and *%ov* indicates the percentage that can be attributed to TCP overhead of total bytes transmitted.

	First Time Retrieval				Cache Validation			
	Pa	Bytes	Sec	%ov	Pa	Bytes	Sec	%ov
HTTP/1.0	510.2	216289	0.97	8.6	374.8	61117	0.78	19.7
HTTP/1.1	281.0	191843	1.25	5.5	133.4	17694	0.89	23.2
HTTP/1.1 Pipelined	181.8	191551	0.68	3.7	32.8	17694	0.54	6.9
HTTP/1.1 Pipelined w. compression	148.8	159654	0.71	3.6	32.6	17687	0.54	6.9

ing. Even so, the Microscape HTML page was compressed by a factor of more than three, shrinking from its original 42 K to 11 K.

This degree of reduction of file size, some aspects of which we've illustrated in Figs. 13-3 and 13-4, proved typical of using the default zlib algorithm on HTML files. Or to put it another way, the demands on bandwidth presented by such files were reduced by about 19 percent after zlib-based compression.

IN A NUTSHELL

1. Web pages should be designed with speed of download as their foremost design criterion.

2. Keeping graphics to a minimum, using multimedia sparingly, and using style sheets are among the techniques that can maintain or

improve the effectiveness of pages without increasing download time.

3. For HTTP/1.1 to outperform HTTP/1.0 in elapsed time, it must include pipelining.

4. Properly buffered pipelined implementations of HTTP/1.1 will improve performance and reduce network traffic.

5. HTTP/1.1 implemented with pipelining outperformed HTTP/1.0 in tests conducted by members of the W3C, even when the HTTP/1.0 implementation used multiple connections in parallel. In terms of packets transmitted, the savings in these tests were at least a factor of two. Elapsed time improvement was less dramatic, but significant.

6. Although the bandwidth savings shown by the researchers from the W3C to be attributable to HTTP/1.1 and its associated TCP-related and connection management techniques are modest, using this version of the protocol in combination with sophisticated, intelligent caching can produce further reductions in Web-related bandwidth consumption.

7. The W3C researchers estimated that if all the techniques they investigated described in this chapter were applied, HTML pages like the Microscape test document might be downloaded over a modem in only about 60 percent of the time needed to download the same document in the same way under HTTP/1.0.

8. The single factor that provided the greatest reduction of bandwidth consumption to HTTP/1.1 was the use of transport compression.

9. Style sheets and image format conversion also provided some bandwidth savings in the W3C tests.

LOOKING AHEAD ■ ■ ■ ■ ■ ■ ■

Chapter 14 examines one last aspect of bandwidth consumption and optimization. It looks at a case study for improving network performance in environments that must integrate a variety of heterogeneous operating system, server, and client software.

14

Integrating Environments

By the year 2000, it's predicted that about 90 percent of all enterprise networking environments will include both UNIX and Windows NT in some fashion. Should this prediction even approach being borne out, businesses and other organizations will be forced to address questions of networking technologies integration.

Among the vendors already seeking to make the UNIX/Windows NT picture a harmonious one is Digital Equipment Corporation. Through its AllConnect for UNIX Program, DEC intends to help organizations clear the hurdle of incorporating both UNIX and Windows NT in their environments.

In this chapter, we examine only one of the areas under investigation by AllConnect: integrating UNIX and Windows NT mail and other messaging. Along the way, we pay particular attention to those aspects of this investigation that present the greatest potential impact on overall bandwidth consumption. By doing so, we hope to offer a model of sorts for bandwidth-efficient integration of any operating system and application suites within a single environment.

NOTE: *On August 2, 1995, DEC and Microsoft announced a formal alliance to provide a complete set of computing products, services, and field support to meet the most demanding needs of customers in large enterprises. This Alliance for Enterprise Computing set the stage for many subsequent activities, including the AllConnect for UNIX Program, an initiative that seeks to provide the technology, tools, services, system integration, and support needed to integrate Windows NT into UNIX-based enterprise computing.*

Integrating Mail and Messaging

Integration of mail and messaging systems across UNIX and Windows NT is the aspect of reconfiguration of a network most likely to affect an entire user population. As such, it must be carefully planned. AllConnect's research leads the project to conclude that nearly all enterprises will manage computer installations including both UNIX and Windows NT for at least the next decade, and probably longer.

The study further cites the impracticality of migrating large numbers of users to Microsoft Exchange at one time. Because some mail systems are embedded in critical applications such as digital signature authorizations or financial transactions, integrating those applications with Microsoft Exchange will take time.

AllConnect also notes that many enterprises have a significant investment in UNIX, as well as in user skills in UNIX-based mail and messaging systems, but at the same time are adopting Windows NT and Microsoft Exchange, and concludes that the two operating system/application suites will coexist for some time. AllConnect therefore recommends a heterogeneous mail system that adheres to industry standards for:

- reliability
- security
- manageability
- efficiency

The Makeup of a Mail System

AllConnect considers any standard mail system to contain:

    I. Servers, which function to collect and forward mail to users and which therefore require:
  A. A message transport protocol specific to each mail messaging system
  B. Server support for user access
  C. A method of properly presenting message formats to users. (For example, in the context of UNIX/Windows NT integration, it is necessary to send a Microsoft Word document to a UNIX workstation in such a manner that the recipient can use the document.)
  D. Directory services, which can range from simple files of user addresses to sophisticated, networkwide database subsystems, and which have two functions:
    1. Storing mail and providing proper mail addresses for all end users
    2. Synchronizing directories across mail servers in order to maintain updated, correct addresses
  E. Connectivity for servers, including:
    1. Transmission media
    2. Networking protocols

    II. A mail client, capable of running on a PC, workstation, or character-based terminal. This client itself has three components:
  A. User agent/interface
  B. Message access protocol
  C. A method for accessing directory services in order to find and use mail system users' addresses

    III. Management utilities

Implications for Bandwidth Consumption of the AllConnect Model. There are several implications to bandwidth consumption of the AllConnect model:

1. It assumes a number of protocols: OSI network, transport, session, and application; mail; directory service; and management.

2. It implies the need to translate document and file formats, as in the Word-to-UNIX example. By so doing, it also implies an additional source of demand for both server and client system resources.

3. In specifying a mail client capable of running on several platforms, it implies the possibility of requiring more than one such application. By so doing, it also implies another source of demand for server resources.

UNIX Mail Systems.   Until recently, UNIX mail systems were centralized mail systems, as opposed to logically distributed client/server implementations, even when the computing environment included UNIX workstations. This means that when UNIX workstation users write and send mail, their mail agent software works on the central server, which provides access to the server's mail application.

NOTE: *As AllConnect points out, the centralized nature of traditional UNIX mail systems is one of the greatest challenges to integration with Windows NT, a completely client/server system.*

Such a centralized UNIX mail system includes:

I. Message transport, through either or both of:
   A. The *simple mail transport protocol* or SMTP, implemented in most versions of UNIX as the sendmail utility; sendmail manages more than just store and forward between mail servers. It also accomplishes server support for user access, by establishing and maintaining a connection between its version on the server and in the mail user's agent
   B. The *post office protocol version 3* or POP3 protocol, which enables downloading of messages to the client desktop

II. Message format conversion, through either:
   A. UUENCODE/UUDECODE, under which the sender uses the first of these utilities to convert mail to an intermediate ASCII format for transmission, while the recipient uses the second to restore the message to its original format
   B. MIME, which automatically converts formats

III. Directory services, either through:
   A. A systemwide database of user names, available to sendmail
   B. User-specific files that define mail aliases, available only to the users who own them

IV. The UNIX message transport protocol SMTP, now the standard protocol for Internet mail

━ ━ ━ ━ ━ ━ ━ ━ ━ ━ ━ ━ ━ ━ ━ ━ ━ ━ ━ ━

NOTE: *AllConnect points out that this broad acceptance of SMTP would permit UNIX SMTP-style mail systems to integrate easily into any NT environment intended to be Internet-capable.*

V. TCP/IP as the network/transport layer suite

VI. A variety of mail clients, depending on the nature of the desktop, some of which rely on SMTP, while others are based on POP3

VII. Message access protocols, which make use of all of sendmail, SMTP, and POP3 in client user agents

VIII. Management utilities, which ordinarily are part of the operating system in UNIX environments but which interact with SMTP

Implications for Bandwidth Consumption of UNIX Mail Systems. By making available more than one message transport protocol, message conversion method, and means of managing directory services, UNIX mail systems present the potential for increased demands on both bandwidth and server resources.

SMTP. As noted in the last section, SMTP, the UNIX mail transport standard, has been accepted by the data communications industry at large as the message transport protocol for the Internet. The term *Internet mail* therefore means *mail stored and forwarded according to the definitions of the SMTP protocol.* As a result, all existing SMTP mail systems can exchange messages across the Internet. Even the increasingly popular mail agents embedded in Web browsers, by relying on SMTP, can be considered independent of their desktop. So, this industry standardization on SMTP as the Internet mail protocol removes at least one source of potential demand for bandwidth.

Implications for Bandwidth Consumption of SMTP. Industrywide standardizing on SMTP as the Internet's mail protocol removes some potential for demands on bandwidth and server resources that might otherwise have been called upon to carry out protocol translation.

Microsoft Exchange Mail. AllConnect considers Microsoft Exchange–based mail systems to include:

I. Mail servers, which are Windows NT systems on either Intel or Alpha processors, running the Microsoft Exchange Server application.

II. Message transport protocol, which may be any of:
   A. RPC: Messages among Exchange Servers typically travel via Microsoft's Remote Procedure Call (RPC) protocol. RPC is particularly useful over local area networks.
   B. SMTP: Microsoft Exchange offers the option of SMTP as a transport protocol, which is sufficient for participation in a UNIX SMTP mail environment. The Exchange Internet Mail Connector product provides SMTP capability to an Exchange server.
   C. X.400:   A third option for a transport protocol is X.400.

III. Message access protocol: Microsoft's *Mail Application Programming Interface* (MAPI) as the client interface for Exchange Server for messaging services from Windows platforms. The MAPI architecture divides messaging applications into three components:
   A. The client application (typically a Windows GUI application)
   B. The MAPI subsystem (provided by Microsoft)
   C. Messaging service providers, to access a chosen messaging server/system
   The MAPI architecture enables developers to create client applications that are easy to maintain and enhance because there is no need to develop code for each messaging system.

IV. Message format conversion, including:
   A. Between Exchange Server and Client, an internal format for converting different file formats between users.
   B. When using X.400 transport, a format standard called the File Transfer Body Part, also supported by MailWorks for UNIX, but not available in the default UNIX mail system or by many X.400 implementations.

V. Directory Services handled by the X.500 international directory service standards.

VI. Address Book, a capability that enables users to use a point-and-click interface to search through other Exchange Server user addresses and choose the appropriate address or addresses to use as a destination.

VII. Microsoft Exchange Client, which allows users to add their own X.400 or SMTP addresses to their personal address books through a provided template.

VIII. The ability to import addresses from an X.400 mail system in bulk into Microsoft Exchange Server and to then merge these addresses into the Exchange Server address list.

IX. Synchronization through the Microsoft Exchange Directory Synchronization Agent, which allows administrators to perform directory synchronization with other mail systems based on Microsoft's MS-Mail Directory Format but which does not provide the ability to allow other X.500-based directories to communicate directly with the Exchange Directory.

X. TCP/IP as the network protocol for its mail and messaging system.

XI. A mail client, which varies according to the desktop, but which can include:
A. Exchange Client.
B. Clients with MAPI Service Provider Interfaces (SPIs).

XII. As a message access protocol between Exchange Client and Exchange Server, either MAPI or POP3.

XIII. As a directory service access protocol, either MAPI or the *Lightweight Directory Access Protocol* (LDAP).

XIV. Management through a dedicated management station for Exchange installations.

Implications for Bandwidth of Microsoft Exchange.　Exchange's:

- reliance on RPC as its primary mail transport protocol

- requirement for a Connector to allow SMTP and RPC to converse

- offering two message access and directory service access protocols

present the possibility of bandwidth overhead beyond that experienced by pure SMTP environments.

Integrating UNIX and Windows NT Mail Systems

In this section, we investigate a number of the issues involved in such an integration.

Message Transport Protocol.   Because SMTP is the standard message transport protocol for UNIX systems, and because Microsoft Exchange uses Microsoft RPC to route mail messages among servers, a common transport protocol must be selected that will be supported by both environments.

In such a scenario, the available options are:

- adding the Internet Connector, an option of Exchange Server, in order to enable communication via SMTP to SMTP mail servers, and thereby effectively making SMTP the mail messaging platform
- using X.400 as the mail messaging protocol, and implementing SMTP and Exchange to connect to it, thereby making the mail transport platform X.400 for both environments
- using both SMTP and X.400

Mail Messaging Backbone.   A *mail messaging backbone,* that is, the intermediate mail server software that must integrate diverse mail systems, must obviously be carefully chosen. Let's examine the alternatives—SMTP, X.400, and X.500—more closely.

X.400.   X.400 has the components:

- *Access Units,* which interface message handling systems to delivery services such as FAX
- *Message Store* (MS), a holding area where messages wait until a recipient chooses to access and read them
- *Message Transfer Agent* or MTA, which stores and forwards messages within and among networks
- *Message Transfer Systems* or MTSes, which are the set of all MTAs
- *User Agents* or UAs, software that creates messages in standardized formats

X.400 standards have been defined for over ten years. As a result, most mail systems provide some form of X.400 support. However, X.400 can be difficult to configure.

X.400 AND SMTP.   Table 14-1 outlines the similarities and the differences between these two mail messaging backbones.

TABLE 14-1	X.400	SMTP
Comparing X.400 and SMTP	*Auditing*: All X.400 implementations provide delivery and nondelivery notes, read receipts, and time and date stamp messages, thus creating audit trails for applications built on mail systems.	*Auditing*: Provided for some SMTP systems, by vendors other than the suppliers of the SMTP systems. Innosoft, for example, supplies deliver receipts only, for various UNIX systems.
	Management: From the time a user sends a message, that message is assigned a unique message identification number that allows it to be tracked from system to system through the network.	*Management*: Messages cannot be uniquely identified, tracked, or managed once sent.
	Reliability: X.400 systems write messages to disk and retain them until the next system in the messaging network acknowledges that it has written them safely to disk using the message identification number and a tracking log.	*Reliability*: Systems may, but often don't, write messages to disk. This can improve the performance of the messaging system, with a trade-off in reliability.
	Administration: X.400 products are separate from the base operating system and must be installed and managed as a separate application.	*Administration*: SMTP—and sometimes MIME—is included with all UNIX vendors' operating environments and is installed as part of the operating system.
	Security: Encryption of messages is automatic.	*Security:* Encryption of messages is a user action via UUENCODE, or, where available, Secure MIME or S/MIME.

X.500. The X.500 information model, although not in the strictest sense a part of a messaging system, relates closely to such systems, since it defines a schema that in turn sets out rules for directory access. X.500 and other directory access service models therefore offer access to such needed information as mail addresses, routing information, and user profiles.

X.500 defines a standard for a distributed directory system, and provides a stock method of representing and identifying users and resources across multivendor, heterogeneous systems. X.500 also defines how information stored in the directory can be accessed.

Developed in 1988 by the ITU, X.500 supports remote directory access, centralized and distributed topologies, and centralized and distributed update methods, as well as peer-entity authentication, digital signatures, and certificates. The standard was updated in 1993 to add support for replicated topologies and access control.

Here's a thumbnail sketch of X.500 operations:

- When creating a message, a user has access to the X.500 directory for address information.

- When transmitting a message, an MTA accesses the X.500 directory for routing information.

- When delivering a message, the MTA checks User Profiles, also housed in the X.500 directory, and which defines the types of documents a user may receive. If the document being sent is in an acceptable format, it is delivered. If not, it is automatically converted to such a format.

X.500 directories often also are used to support tasks such as:

- housing personnel information

- housing customer information

- housing security and cryptographic information

- serving as the basis for White and Yellow Pages

Message Access Protocol. Message access protocols become an issue when UNIX and Windows NT:

- must be configured in a variety of combinations of server and client

- must match client mail user agents with servers that support both a user's desktop device preference and preferred style of mail manipulation

In such scenarios, three choices exist.

- post office protocol version 3, or POP3

- the interactive mail access protocol (IMAP)

- proprietary protocols

POP3 OR IMAP? Depending on the user and client involved, the perceived relative strengths and weaknesses of these protocols may vary. Table 14-2 summarizes the differences and similarities between these two most widely used message access protocols.

TABLE 14-2

Comparing POP3
and IMAP4

In This Area	POP3 Offers	IMAP4 Offers
Mail access	Off-line, on-line, and disconnected access	Off-line, on-line, and disconnected access
Folders	Single	Multiple
Centralized mail folders?	No	Yes
Folder sharing by users?	No	Yes
Folder manipulation?	None	Create, delete, append, move, copy
Search capability?	None	Search of headers and body
Message manipulation	Delete	Delete, move, copy, modify

Implementation Suggestions

We close this chapter by ticking off a list of AllConnect's most significant suggestions for integrating UNIX and Windows NT mail services.

MIME. Make all UNIX systems MIME capable. For users with existing UNIX mail systems, a first step is to upgrade all UNIX mail servers
to include MIME support. A number of systems offer such support,
among them DIGITAL UNIX Version 4.1 and later, the HP OpenMail
user agent, and Sun Microsystems' Mail Tool.

POP3 clients provide MIME support, even if their server doesn't. Similarly, MailWorks for UNIX provides MIME support, even to legacy
clients like cc:Mail and MSMail.

Connectivity Patterns. Use point-to-point connectivity. Sites with
three or fewer diverse mail systems, and fewer than 30 total systems will
most likely find a point-to-point connection between UNIX and
Microsoft Exchange to be sufficient. Such sites will therefore also need to
use the Internet Connector product for Exchange in order to integrate
Exchange with SMTP as well as with the Internet.

Mail Services Backbone. Use an X.400 backbone, since this standard:

- bridges multiple mail systems, not only UNIX and Windows NT, but also such OSes and environments as OpenVMS and mainframes
- increases manageability
- provides for easy growth

Directory Services. Use X.500 directory services; this model offers much the same benefits as X.400.

IN A NUTSHELL

In this section, we recap the implications for bandwidth consumption of the need to integrate differing application environments, such as UNIX and Windows NT mail services.

1. Integration of application environments frequently necessitates a number of protocols: OSI network, transport, session, and application; mail; directory service; and management.

2. Integration implies the need to translate document and file formats. It therefore also implies an additional source of demand for both server and client system resources.

3. If it assumes or offers clients capable of running on several platforms or in several desktop operating environments, integration implies the need for multiple protocols or for protocol translation or connection, as well as higher levels of demand for server resources.

4. By making available more than one message transport protocol, message conversion method, and means of managing directory services, UNIX mail systems present the potential for increased demands on both bandwidth and server resources.

5. Exchange's:
   - reliance on RPC as its primary mail transport protocol
   - requirement for a Connector to allow SMTP and RPC to converse
   - offering two message access and directory service access protocols

   present the possibility of bandwidth overhead beyond that experienced by pure SMTP environments

6. Industrywide standardizing on SMTP as the Internet's mail protocol removes some potential for demands on bandwidth, and on
server resources that might otherwise have been called upon to
carry out protocol translation.

LOOKING AHEAD

No doubt you've come to the conclusion that maximizing bandwidth
requires close monitoring of a network. In the next chapter, we examine
nearly four dozen tools that allow you to do just that.

CHAPTER 15

Network
Monitoring Tools

In this chapter, we present synopses of nearly four dozen tools for monitoring network performance. The catalog from which the format for this chapter was drawn was compiled by members of the *Internet Engineering Task Force* (IETF). Like the members of the IETF who created much of the source material for this chapter, we have not extensively tested these tools. But we have verified the tools' characteristics with their creators.

Once again like the IETF's NOCTools Working Group, we present each tool's summary in the following format:

- Tool Name
- Network aspects to which the tool pertains
- Abstract (a brief description of the tool)
- Mechanism (how the tool does its job)
- Caveats (any cautions to be aware of in using the tool)

- Bugs (assuming there are any; few of the tools described in this chapter have them)
- Limitations (ditto)
- Hardware Required, if any
- Software Required, if any
- Availability of the tool

The Tools

In this section, we offer descriptions of 44 tools for managing TCP/IP networks. The tools are presented in alphabetical order by name.

Application Development Toolkit

Pertains to general network management.

Abstract. snmpapi is a toolkit for developing SNMP applications and agents, and is well suited for embedded systems such as bridges or routers. An example, MIB II agent for Sun Sparcstations, is provided; snmpapi is distributed in source form only.

Mechanism. snmpapi is a library of C functions.

Caveats. None.

Bugs. None known.

Limitations. None known.

Hardware Required. No restrictions.

Availability. Available now. For more information, go to *http://www.avatar.com/products.htm.*

arpwatch

Pertains to Ethernet and IP.

Abstract. arpwatch is an Ethernet monitor program that keeps track of Ethernet/ip address pairings. This package requires the bpf and libpcap packages.

Mechanism. arpwatch, a tool that monitors Ethernet activity and keeps a database of Ethernet/ip address pairings, also reports certain changes via email. arpwatch uses libcap, a system-independent interface for user-level packet capture. Before building this tool, you must first retrieve and build libpcap. Once libpcap is built (either install it or make sure it's in ../libpcap), you can build arpwatch.

Caveats. None.

Bugs. None known.

Limitations. None known.

Hardware Required. No restrictions.

Software Required. UNIX.

Availability. Available, though subject to copyright restrictions, via anonymous FTP from *ftp.ee.lbl.gov*, an FTP server of the Network Research Group of the Lawrence Berkley Laboratory. The source and documentation for the tool is in compressed tar format, in the file *libp-cap-*.tar.Z*.

AUTONET/Performance

Pertains to network traffic and status.

Abstract. AUTONET/Performance is a network performance analysis tool for WANs and LANs. It analyzes the performance of the distribution and backbone network to determine response times and standard deviation.

Mechanism. AUTONET/Performance is a network performance analysis tool for WANs and LANs. It will analyze the performance of the distribution and backbone network to determine response times, standard deviation, input and output queuing delays, and I/O line utilization. AUTONET/Performance will perform Sensitivity Analysis to determine performance under varying traffic conditions, and Capacity Analysis to determine the maximum traffic the network can handle given response time and line utilization constraints.

Caveats.   None.

Bugs.   None known.

Limitations.   Source code is not available.

Hardware Required.   Sun SPARC.

Software Required.   Solaris Sparc 1.0, 2.0.

Availability.   AUTONET/Performance is a commercial product of Network Design & Analysis Corporation, 60 Gough Road Suite 2A, Markham, Canada, L3R 8X7. Phone: (416) 477-9534.

Chameleon 20 and 22

Pertains to   network traffic and traffic generation; overall network status.

Abstract.   Tekelec's ChameLAN 100 is a portable diagnostic system for monitoring and simulation of FDDI, Ethernet, and Token Ring networks—simultaneously. Protocol analysis of multiple topologies, as well as mixed topologies simultaneously, is a key feature of the product family. Tekelec's proprietary FDDI hardware guarantees complete real-time analysis of networks and network components at the full ring bandwidth of 125 Mbps. The hardware connects to the network and captures 100 percent of network data, measures performance, and isolates real-time problems.

The simulation option offers full bandwidth load generation that allows you to create and simulate any network condition.

Monitoring of FDDI, Ethernet, and Token Ring allows the user to view network status in real time; view network, node, or node pair statistics; capture frames; control capture using trigger and filter capabilities;

view real-time statistics; view captured frames in decoded format; and view the last frame transmitted by each station.

Mechanism.   Standard graphical Motif/X-Windows and TCP/IP allow remote control through Ethernet and 10 Base T interfaces.

Caveats.   None.

Bugs.   None known.

Limitations.   None known.

Hardware Required.   None. These are self-contained units, with their own interface cards.

Software Required.   None.

Availability.   Available now. See *http://webserver.tekelec.de/com/ch20_22e.htm* for more information.

Chariot

Pertains to   network traffic and status.

Abstract.   Chariot, as illustrated in Fig. 15-1, is a test tool that determines end-to-end performance of complex, multiprotocol networks.

Mechanism.   Chariot is a software-based performance tool that will help you:

- evaluate the performance and capacity of almost any network device, including routers, switches, adapters, and network software
- stress or regression test network equipment
- provide the measurements you need to help tune your network
- anticipate the effects of rolling out new applications
- determine the performance you're getting from network services, such as a Frame Relay network

Figure 15-1
Measurements such as those Chariot can offer are critical to optimizing bandwidth use.

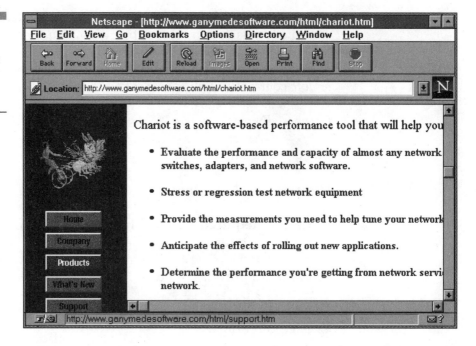

There are two components in Chariot: a console and endpoints. You create and run tests at the console. The console distributes scripts to the endpoints, and the endpoints execute the instructions in the script. Each endpoint is a network application that understands how to execute many different test scripts over different protocols. When an endpoint runs a script, it looks just like an application to the network. Each endpoint can support multiple connections, run different scripts, and mix protocols. When a test completes, the Chariot console collects the results from the endpoints and generates a report with the information you need to know about your application: response time, transaction rates, and throughput.

Caveats.   None.

Bugs.   None known.

Limitations.   None known.

Hardware Required.   Chariot runs as an application on the communication stack. Therefore, it can run over any DLC that is supported by the stack you are using. The hardware requirements for Chariot depend on the operating system and protocols in use, and the size and purpose of the test. Chariot has run on everything from 386 PCs to Pentiums loaded with memory.

Software Required.   The range of software platforms on which Chariot will run is extensive, to say the least. Table 15-1 gives only a selection of software platforms for the Chariot endpoint.

Availability.   Chariot is a commercial product of Ganymede Software Inc. For information on this tool, visit the Ganymede Web site, shown in Fig. 15-2, at *http://www.ganymedesoftware.com*.

CMU SNMP Distribution

Pertains to   general network management and status.

Abstract.   The CMU SNMP Distribution includes source code for an SNMP agent, several SNMP client applications, an ASN.1 library, and supporting documentation.

   The agent compiles into about 10 KB of 68000 code. The distribution includes a full agent that runs on a Kinetics FastPath2/3/4, and is built into the KIP appletalk/ethernet gateway. The machine independent portions of this agent also run on CMU's IBM PC/AT-based router. The applications are designed to be useful in the real world.

Mechanism.   SNMP.

Caveats.   None.

Bugs.   None known.

Limitations.   None known.

Hardware Required.   The gateway agent runs on a Kinetics FastPath2/3/4. Otherwise, no restrictions.

Software Required.   The code was written with efficiency and portability in mind. The applications compile and run on the following sys-

TABLE 15-1 Summarizing Chariot

This Version of Chariot	Needs This Operating System	This TCP/IP Stack	This SNA/APPC Stack	This IPX/SPX Stack
Windows 3.1 Endpoint	Windows 3.1, Windows for Workgroups 3.11	A variety of TCP/IP stacks are supported	Not supported	Not supported
Windows 95 Endpoint	Windows 95	TCP/IP included with Windows 95, with MS Service Pack 1	IBM Personal Communications v4.11 for Windows 95	Novell Client32 for Windows 95; or Microsoft Windows 95B
Windows NT for x86 Endpoint	Windows NT versions 4.0 or 3.51, workstation or server	TCP/IP (WinSock) stack included with Windows NT Windows NT v4.0 Service Pack 3 is recommended	Microsoft SNA Server v3.0 or v2.11 with Service Pack 2; or IBM Communications Server for Windows NT v1.0; or IBM Personal Communications for Windows NT v4.11	IPX/SPX stack included with Windows NT
Windows NT for Alpha Endpoint	Microsoft Windows NT for Alpha v4.0, workstation or server	TCP/IP (WinSock) stack included with Windows NT	Microsoft SNA Server for Alpha v2.11 with Service Pack 1	IPX/SPX stack included with Windows NT
OS/2 Endpoint	IBM OS/2 Warp 4; or OS/2 Warp Connect 3; or OS/2 v2.11	OS/2 Warp 4 or Warp Connect 3 or IBM TCP/IP for OS/2 2.0	IBM Communications Server for OS/2 v4.1; or IBM Communications Manager/2 v1.11	Novell's Client for OS/2 version 2.12
Novell NetWare Endpoint	Novell NetWare v4.x or 3.12	TCP/IP stack included with NetWare	Not supported	IPX/SPX stack included with NetWare

AIX Endpoint	AIX (RS/6000) version 4.1	TCP/IP included with AIX	Not supported	Not supported
Digital UNIX Endpoint	Digital UNIX for Alpha v4.0 or later	TCP/IP stack included with Digital UNIX	Not supported	Not supported
HP-UX Endpoint	HP-UX version 10.0 or 9.0	TCP/IP included with HP-UX	Not supported	Not supported
Linux Endpoint	A Linux v2.0 system, such as Slackware (kernel 2.0.0) or Red Hat (kernel 2.0.27)	TCP/IP stack included with Linux	Not supported	Not supported
Sun Solaris Endpoint	Sun Solaris version 2.4	TCP/IP included with Sun Solaris	Not supported	Not supported

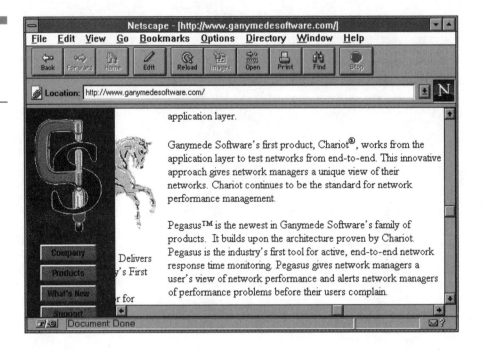

tems: IBM PC/RT running ACIS Release 3, Sun3/50 running SUNOS 3.5, and the DEC microVax running Ultrix 2.2. They are expected to run on any system with a Berkeley socket interface.

Availability. This distribution is copyrighted by CMU, but may be used and sold without permission. Consult the copyright notices for further information. The distribution is available by anonymous FTP from the host *lancaster.andrew.cmu.edu* (128.2.13.21) as the files *pub/cmu-snmp.9.tar*, and *pub/kip-snmp.9.tar*. The former includes the libraries and the applications, and the latter is the KIP SNMP agent.

DCE/Sleuth

Pertains to network traffic and status.

Abstract. DCE/Sleuth, depicted in Fig. 15-3, is a DCE (Open Software Foundation Distributed Computing Environment) network analyzer.

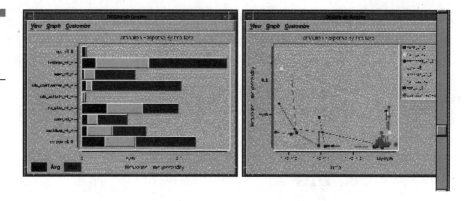

Figure 15-3
DCE/Sleuth provides a variety of display formats.

Mechanism. DCE/Sleuth features include:

- full DCE protocol decoding
- user-programmable IDL interpretation
- multilevel decoding
- advanced filtering, by host or protocol/port
- multiple graphs and reports, including response time, RPC count, packet count, and byte count

DCE/Sleuth speeds resolution of DCE-related networking problems, and provides a means of service-level monitoring by timing RPC transactions.

Caveats. None.

Bugs. None known.

Limitations. None known.

Hardware Required. Memory: 32 MB (min.), 48 MB (preferred); storage: 5 MB (installation).

Software Required. DCE/Sleuth will run under a number of versions of UNIX, including:

- Solaris 2.4/2.5
- AIX 4.x

- HP/UX 10.x
- Digital UNIX 4.x
- SINIX 5.4

Availability.   DCE/Sleuth is a commercial product of Intellisoft Corporation. For more information on DCE/Sleuth, visit the Intellisoft Web site at *http://www.isoft.com.*

DiG

Pertains to   general network and DNS status.

Abstract.   DiG, which stands for *domain information groper,* is a command line tool that queries DNS servers in either an interactive or a batch mode. It was developed to be more convenient/flexible than nslookup for gathering performance data and testing DNS servers.

Mechanism.   DiG is built on a slightly modified version of the bind resolver (release 4.8).

Caveats.   None.

Bugs.   None known.

Limitations.   None known.

Hardware Required.   No restrictions.

Software Required.   BSD UNIX.

Availability.   DiG is available via anonymous FTP from *venera.isi.edu* in the file *pub/dig.2.0.tar.Z.*

EMANATE: Enhanced MANagement Agent Through Extensions

Pertains to   general network management and status.

Abstract. EMANATE provides a run-time extensible SNMP agent that can support zero, one, or many subagents. EMANATE consists of several logically independent components and subsystems, including:

- a Master SNMP agent, which contains an API to communicate with subagents
- a Subagent Developer's Kit, which contains tools to assist in the implementation of subagents
- EMANATE libraries, which provide the API for subagents

Mechanism. A concise API allows a standard means of communication between the master and subagents. System dependent mechanisms are employed for transfer of information between the master and subagents.

Caveats. None.

Bugs. None known.

Limitations. None known.

Hardware Required. Multiple platforms including PCs, workstations, hosts, and servers are supported.

Software Required. C compiler.

Availability. EMANATE is a commercial product available under license from SNMP Research. For more information, see *http://www.snmp.com.*

Generic Managed System

Pertains to general network management.

Abstract. The *Generic Managed System* or GMS implements a number of common OSI network management functions, including the parsing of requests, selection of managed objects, and handling of notifications. The intention is that implementors should use GMS as a basis for their own managed object implementations.

Mechanism.   GMS uses the UCL CMIP library plus a library of C++ objects representing common managed objects and attribute types.

Caveats.   The system is under ongoing development and enhancement, and is not well documented.

Bugs.   None known.

Limitations.   None known.

Hardware Required.   Has been tested on SUN 3 and SUN 4 architectures.

Software Required.   The ISODE protocol suite, BSD UNIX, UCL CMIP Library, GNU C++ (g++).

Availability.   The CMIP library and related management tools built upon it, known as *OSIMIS* (OSI Management Information Service), are publicly available from University College London, England, via FTP and FTAM. To obtain more information on this tool, see *http://www.cs.ucl.ac.uk/research/report/JCNETMAN.htm*. For information regarding obtaining the tool, send email to *osimis-request@cs.ucl.ac.uk* or call + 44 71 380 7366.

Internet Rover

Pertains to   general network status; IP, SMTP.

Abstract.   Internet Rover is a network monitor that uses multiple protocol samples to test network health. The package consists of two modules: a data collector and a problem display. Internet Rover includes only one data collector, which carries out all network tests and maintains a list of problems on the network. However, the Rover offers the option of running multiple display processes simultaneously.

The display module uses the UNIX utility curses, allowing many terminal types to display the problem file either locally or from a remote site. The data collector is easily configured and extensible.

Mechanism.   Internet Rover relies on a configuration file that contains a list of nodes, addresses, a NodeUp? protocol test (usually, a simple ping), and a list of further tests to be performed if the node is in fact up. Test modules include ones for TELNET, FTP, and SMTP.

Caveats.   None.

Bugs.   None known.

Limitations.   Internet Rover finds and displays, but doesn't correct, problems on a network.

Hardware Required.   This software has been run on Sun workstations and IBM Remote Terminals.

Software Required.   4.x BSD UNIX socket programming libraries, BSD ping, curses.

Availability.   Full source code for Internet Rover can be obtained by anonymous FTP from merit.edu in the directory *ftp.merit.edu/internet.tools/rover.* Both source and executables are public domain and can be freely distributed for noncommercial use.

LanProbe

Pertains to   general network status and traffic.

Abstract.   LanProbe is a distributed monitoring application suite, which carries out remote and local monitoring of Ethernet LANs and is protocol- and vendor-independent. LanProbe discovers each active node on a segment and displays that segment on a network map, along with information including the segment's adapter card vendor, Ethernet address, and IP address.

When its NodeLocator option is used, LanProbe automatically gathers data about the location of existing or newly configured nodes, thereby making its map an up-to-date picture of the physical layout of the segment.

LanProbe also can gather and display traffic statistics that can be exported for further analysis. The application also accepts user-defined thresholds as the basis for automating alerts.

LanProbe's trace module provides both a remote protocol analyzer and decoding of common protocols.

Other significant but not strictly protocol-related network events, such as power failures or cable breaks are tracked in a log that is automatically uploaded to LanProbe's View module periodically. That module, actually called *ProbeView*, generates reports that can be manipulated by word processors, spreadsheets, and DBMS.

Mechanism. The system consists of one or more LanProbe segment monitors and ProbeView software. The LanProbe segment monitor attaches to the end of an Ethernet segment and monitors all traffic. Attachment can be direct to a thin or thick coax cable, or via an external transceiver to fiber optic or twisted pair cabling. Network data relating to the segment is transferred to a workstation running ProbeView via RS-232, Ethernet, or a modem connection.

ProbeView software, which runs on a PC/AT class workstation, presents network information in graphical displays.

The HP4992A NodeLocator option attaches to the opposite end of the cable from the HP4991A LanProbe segment monitor. It automatically locates the position of nodes on the Ethernet networks using coaxial cabling schemes.

Caveats.   None.

Bugs.   None known.

Limitations.   None known.

Hardware Required.   HP 4991A LanProbe segment monitor, HP 4992A NodeLocator (for optional capabilities).

Software Required.   HP 4990A ProbeView.

Availability.   A commercial product available from Hewlett-Packard Company. For details on availability, see the H-P Web site at *http://www.tmo.hp.com/tmo/ntd/products/probes/probes-quad.html.*

LANWatch

Pertains to   network traffic; DECnet; DNS; IP, SMTP.

Abstract. LANWatch 2.0 is a flexible network analyzer that runs under DOS on personal computers and requires no hardware modifications to either the host or the network. LANWatch is a software-only package that installs easily in existing PCs.

Mechanism. LANWatch uses common PC network interfaces, placing them in promiscuous mode and capturing traffic.

Caveats. Most PC network interfaces will not capture 100 percent of the traffic on a fully loaded network (primarily missing back-to-back packets).

Bugs. None known.

Limitations. The inability to capture back-to-back packets.

Hardware Required. LANWatch requires a PC or PS/2 with a supported network interface card.

Software Required. LANWatch runs in DOS. Modification of the supplied source code or creation of additional filters and parsers requires Microsoft C 5.1.

Availability. LANWatch is commercially available from FTP Software. For more information, see *http://www.ftp.com*.

libcap

Pertains to packet capture.

Abstract. The libpcap packet capture library provides a uniform library interface to various packet capture systems including the Berkeley Packet Filter (BSD and SunOS), Data Link Provider Interface (Solaris and SYSV), Stanford Enetfilter (IBM RT/4.3BSD), Network Interface Tap (SunOS 3), Streams Network Interface Tap (SunOS 4), Packet Filter, and Snoop (IRIX).

Mechanism. See preceding Abstract.

Caveats.   None.

Bugs.   None known.

Limitations.   None known.

Hardware Required.   Any Ultrix system (VAX or DEC RISC hardware).

Software Required.   Ultrix release 4.0 or later. For Ultrix 4.1, may require the patched *if_ln.o* kernel module, available from Digital's Customer Support Center.

Availability.   Available, though subject to copyright restrictions, via anonymous FTP from *ftp.ee.lbl.gov,* an FTP server of the Network Research Group of the Lawrence Berkley Laboratory. The source and documentation for the tool is in compressed tar format, in the file *libcap.tar.z.*

MONET

Pertains to   network topology; routing; network status; network traffic; bridging; alarm management; DECnet; IP; SNMP; UNIX.

Abstract.   MONET provides the capability to manage and control SNMP-based networking products from any vendor including those from Hughes LAN Systems. A comprehensive relational database manages the data and ensures easy access and control of resources throughout the network.

MONET provides multivendor management through its advanced Mib master MIB parser that allows the parsing of enterprise MIBs directly into the RDBMS for use by MONET's applications.

Major features include:

- *Remote access with X.*   Use of the X/Motif user-interface, enabling remote access to all applications.

- *Database management.*   Stores and retrieves the information required to administer and configure the network.

- *Graphics and network mapping.*   The Graphics module enables the user to view the nodes in the network as "dynamic" icons in hierarchical maps.

- *Configuration management.* Retrieves configuration information from SNMP devices.

- *Performance management.* Displays local network traffic graphically, by packet size, protocol, network utilization, sources and destinations of packets, and so forth. Provides for the scheduling of jobs to retrieve MIB values of a device and store them in the RDBMS for review or summary reporting at a later time. Allows high/low thresholds to be set on retrieved values with alarms generated when thresholds are exceeded.

- *Fault management.* Provides availability monitoring and indicates potential problems. Creates alarms from received SNMP traps, and from other internally generated conditions. Records alarms in the alarm log in the RDBMS. Lists alarms for selected set of devices, according to various filter conditions.

Mechanism. SNMP.

Caveats. None known.

Bugs. None known.

Limitations. Maximum number of nodes that can be monitored is 18,000. This can include hosts, terminal servers, PCs, routers, and bridges.

Hardware Required. The host for the NMC software is a Sun 4 desktop workstation. Recommended minimum hardware is the Sun IPX Color workstation, with a $1/4$" SCSI tape drive.

Software Required. MONET V5.0, which is provided on $1/4$" tape format, runs on the Sun 4.1.1 Operating System.

Availability. A commercial product of Hughes LAN Systems Inc., 1225 Charleston Road Mountain View, CA 94043. Phone: (415) 966-7300. Fax: (415) 960-3738.

Net.Medic Pro

Pertains to network traffic and network application performance.

Net.Medic

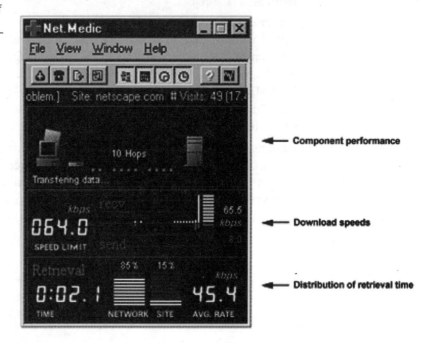

Abstract.   Net.Medic Pro, one aspect of which is illustrated in Fig. 15-4, lets you continuously monitor the ongoing performance of each of your Inter/intranet components, as well as how they work together.

Mechanism.   Net.Medic Pro's AutoMonitor feature automatically monitors application performance over LAN, WAN, and dial-up connections, allowing you to view network transactions as they occur, just as your end users experience them. For example, you can automatically monitor the availability and response times of your Web servers by typing in the URLs and defining the testing intervals. AutoMonitor automatically accesses the servers according to the schedule you define. The Health Log then summarizes the data about any problems encountered during testing, along with the regular, ongoing data collected by Net.Medic Pro. In addition, AutoMonitor can be used to validate service-level agreements (SLAs) with your customer base.

Net.Medic Pro's Notify feature alerts you when a performance-affecting problem occurs during an automated test sequence. An email mes-

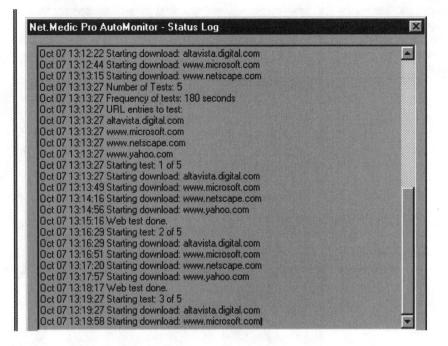

Figure 15-5
The Status Log offers extensive information.

sage is immediately generated, enabling you to proactively respond and minimize the impact on your customers. Net.Medic Pro's Status Log provides underlying details of the problem and its causes, giving you the information you need to take corrective action. The Status Log contains a detailed record of all testing transactions, enabling you to review the details and progress of AutoMonitor tests.

Net.Medic Pro's history reports highlight chronic bottlenecks, peak usage periods, and network trends, so you can proactively meet customers' needs. The Status Log, shown in Fig. 15-5, and history reports provide you with comprehensive data about all aspects of networked application performance.

The AutoMonitor, part of whose configuration Fig. 15-6 illustrates, can also be used to regularly dial into modem banks to test availability and performance. Detailed data such as log-in times, time to connect, connection speeds, busy signals, and other call failures lets you build a performance history as well as track service levels, ingress and egress speeds, and network throughput from the customer's perspective.

Figure 15-6
Net.Medic provides
several configuration
options.

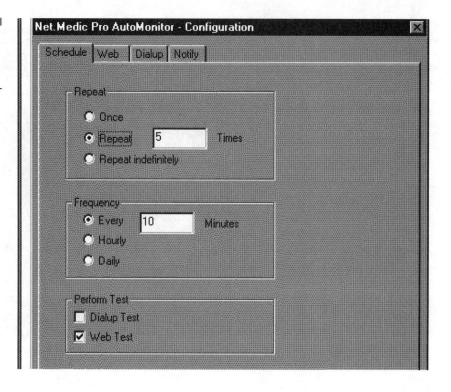

Caveats. None.

Bugs. None known.

Limitations. None known.

Hardware Required. 16 MB RAM or more; 2 MB free disk space or more.

Software Required. Windows 95 or Windows NT 4.0 operating system; Netscape Navigator 3.x and Microsoft Internet Explorer 3.x or higher.

Availability. Net.Medic is a commercial product of VitalSigns Software. Learn more about this and other similar VitalSigns products at their Web site, at *http://www.vitalsigns.com/products.*

NetMetrix Load Monitor

Pertains to alarms; network traffic; FDDI; IP; UNIX.

Abstract. The NetMetrix Load Monitor is a distributed client-server monitoring tool for Ethernet, Token Ring, and FDDI networks. A unique "dual" architecture provides compatibility with both RMON and X windows. RMON allows interoperability and an enterprise-wide view, whereas X windows enables much more powerful, intelligent applications at remote segments and saves network bandwidth.

The Load Monitor provides extensive traffic statistics. It looks at load by time interval, source node, destination node, application, protocol, or packet size. A powerful ZOOM feature allows extensive correlational analysis that is displayed in a wide variety of graphs and tables.

A floating license allows easy access to the software tool anywhere you need it.

Mechanism. NetMetrix turns the network interface into promiscuous mode to capture packets.

Caveats. Originally developed on SPARCStations under SunOS, the NetMetrix family of tools have been acquired by and are now offered in an HP Open View environment.

Bugs. None known.

Limitations. None known.

Hardware Required. See preceding Caveats.

Software Required. See preceding Caveats.

Availability. The NetMetrix family of network monitoring tools is available from Metrix Network Systems, Inc., a subsidiary of Hewlett-Packard. For more information on this and other NetMetrix tools, check the H-P Web site at *http://www.tmo.hp.com/tmo/ntd/*.

NetMetrix NFS Monitor

Pertains to network traffic; FDDI; NFS; UNIX.

Abstract.   The NetMetrix NFS Monitor is a distributed network monitoring tool, which monitors and graphs NFS load, response time, retransmits, rejects and errors by server, client, NFS procedure, or time interval. Breakdown server activity by file system and client activity by user.

A ZOOM feature lets you correlate monitoring variables. You can see client/server relationships, compare server performance, evaluate NFS performance enhancement strategies.

Mechanism.   NetMetrix turns the network interface into promiscuous mode to capture packets.

Caveats.   Originally developed on SPARCStations under SunOS, the NetMetrix family of tools have been acquired by and are now offered in an HP Open View environment.

Bugs.   None known.

Limitations.   None known.

Hardware Required.   See preceding Caveats.

Software Required.   See preceding Caveats.

Availability.   The NetMetrix family of network monitoring tools is available from Metrix Network Systems, Inc., a subsidiary of Hewlett-Packard. For more information on this and other NetMetrix tools, check the H-P Web site at *http://www.tmo.hp.com/tmo/ntd/*.

NetMetrix Protocol Analyzer

Pertains to   network traffic; DECnet, DNS, Ethernet, FDDI, IP, NFS, SMTP; UNIX.

Abstract.   The NetMetrix Protocol Analyzer is a distributed client-server monitoring tool for Ethernet, Token Ring, and FDDI networks. Its architecture provides compatibility with both RMON and X windows. RMON allows interoperability, whereas X windows enables intelligent applications at remote segments and saves network bandwidth.

With the Protocol Analyzer, you can decode and display packets as they are being captured. Extensive filters let you sift through packets either before or after trace capture. The capture filter may be specified by source, destination between hosts, protocol, packet size, pattern match, or by a complete expression using an extensive filter expression language.

Packet decodes are available for all major protocols including DECnet, Appletalk, Novell, XNS, SNA, BANYAN, OSI, and TCP/IP. The decodes for the TCP/IP stack have all major protocols including NFS, YP, DNS, SNMP, OSPF, and so forth.

Request and reply packets are matched. Packets can be displayed in summary, detail, or hex, with multiple views to see packet dialogues side by side.

A complete developers' kit is available to build custom decodes.

Mechanism. NetMetrix turns the network interface into promiscuous mode to capture packets.

Caveats. Originally developed on SPARCStations under SunOS, the NetMetrix family of tools have been acquired by and are now offered in an HP Open View environment.

Bugs. None known.

Limitations. None.

Hardware Required. See preceding Caveats.

Software Required. See preceding Caveats.

Availability. The NetMetrix family of network monitoring tools is available from Metrix Network Systems, Inc., a subsidiary of Hewlett-Packard. For more information on this and other NetMetrix tools, check the H-P Web site at *http://www.tmo.hp.com/tmo/ntd/.*

NetMetrix Traffic Generator

Pertains to network traffic; FDDI, IP, UNIX.

Abstract. The NetMetrix Traffic Generator is a distributed software tool, which allows you to simulate network load or test packet dialogues between nodes on your Ethernet, Token Ring, or FDDI segments. The Traffic Generator can also be used to test and validate management station alarms, routers, bridges, hubs, and so forth.

An easy-to-use programming interface provides flexibility over variables such as bandwidth, packet sequence, and conditional responses.

Mechanism.   NetMetrix turns the network interface into promiscuous mode to capture packets.

Caveats.   Originally developed on SPARCStations under SunOS, the NetMetrix family of tools have been acquired by and are now offered in an HP Open View environment.

Bugs.   None known.

Limitations.   None.

Hardware Required.   See preceding Caveats.

Software Required.   See preceding Caveats.

Availability.   The NetMetrix family of network monitoring tools is available from Metrix Network Systems, Inc., a subsidiary of Hewlett-Packard. For more information on this and other NetMetrix tools, check the H-P Web site at *http://www.tmo.hp.com/tmo/ntd/*.

NETMON for Windows

Pertains to   routing; DECnet, Ethernet, IP, NMS.

Abstract.   NETMON implements a network management station based on a low-cost DOS platform. NETMON's network management tools for configuration, performance, security, and fault management have been used successfully with a wide assortment of wide- and local-area-network topologies and medias. Multiprotocol devices are supported including those using TCP/IP and DECNet protocols.

NETMON's features include:

- Fault management tool: displays a map of the network configuration with node and link state indicated in one of several colors to indicate current status

- Configuration management tool: may be used to edit the network management information base stored in the NMS to reflect changes occurring in the network
- Graphs and tabular tools: for use in fault and performance management
- Mechanisms by which additional variables, such as vendor-specific variables, may be added
- Alarms: may be enabled to alert the operator of events occurring on the network
- Events logged to disk
- Output data: may be transferred via flat files for additional report generation by a variety of statistical packages

The NETMON application comes complete with source code including a set of portable libraries for generating and parsing SNMP messages.

Mechanism.   Based on SNMP. Polling is performed via the SNMP get-next operator and the SNMP get operator. Trap directed polling is used to regulate the focus and intensity of the polling.

Caveats.   None.

Bugs.   None known.

Limitations.   None known.

Hardware Required.   The minimum system is an IBM 386 computer, or compatible, with hard disk drive.

Software Required.   DOS 5.0 or later, Windows 3.0 in 386 mode or later, and TCP/IP kernel software from FTP Software.

Availability.   This is a commercial product available under license from SNMP Research. For more information, visit their Web site at *http://www.snmp.com.*

Netperf

Pertains to   network traffic and throughput.

Abstract.   Netperf, depicted in Fig. 15-7, is a benchmark that can be used to measure various aspects of networking performance. Its primary focus is on bulk data transfer and request/response performance using either TCP or UDP and the Berkeley Sockets interface. There are optional tests available to measure the performance of DLPI, Unix Domain Sockets, the Fore ATM API, and the HP HiPPI LLA interface.

   This tool is maintained and informally supported by the IND Networking Performance Team. It is *not* supported via any of the normal Hewlett-Packard support channels.

Mechanism.   Netperf is designed around the basic client-server model. There are two executables: netperf and netserver. Generally you will execute only the netperf program; the netserver program will be invoked by the other system's inetd.

Figure 15-7
Here's a glimpse at some of what's happening with Netperf.

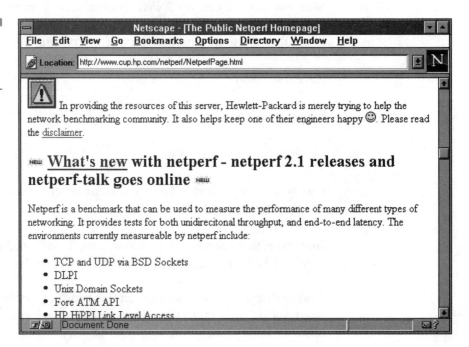

When you execute netperf, the first thing that will happen is the establishment of a control connection to the remote system. This connection will be used to pass test configuration information and results to and from the remote system. Regardless of the type of test being run, the control connection will be a TCP connection using BSD sockets.

Once the control connection is up and the configuration information has been passed, a separate connection will be opened for the measurement itself using the APIs and protocols appropriate for the test. The test will be performed and the results will be displayed.

Netperf places no traffic on the control connection while a test is in progress. Certain TCP options, such as SO_KEEPALIVE, if set as your system's default, may put packets out on the control connection.

CPU utilization is a frequently requested metric of networking performance. Unfortunately, it can also be one of the most difficult metrics to measure accurately. Netperf is designed to use one of several (perhaps platform-dependent) CPU utilization measurement schemes. Depending on the CPU utilization measurement technique used, a unique single-letter code will be included in the CPU portion of the test banner for both the local and remote systems.

The default CPU measurement technique is based on pstat (-DPSTAT compilation only). This technique should work on most HP-UX systems, but it may underreport the CPU usage. The extent of this underreporting is not currently known. When pstat( ) is used to gather CPU utilization information, a "P" will be displayed in the test banner in the CPU column.

A second measurement technique is based on a counter inserted into the HP-UX kernel's idle loop. Whenever the system goes completely idle (from the kernel's perspective), this counter starts to increment. When the system is not idle, the counter stops incrementing. This counter's value is retrieved by reading from /dev/kmem, a process that generally requires superuser privileges. CPU utilization is determined by comparing the rate at which the counter increments when the system is idle to the rate when a test is running. The idle rate is recomputed for each test unless provided by the user.

This counter is not present in a production HP-UX kernel and must be compiled-in. Briefly, this entails adding a flag (-DIDLE_CNT) to the kernel makefile, removing a .o file, and recompiling the kernel. This cannot be done on a system lacking kernel sources. Furthermore, this technique cannot be used at all on non-HP systems unless the vendors in question implement a similar technique. The kernel idle counter is considered highly accurate. When it is used, an 'I' will be displayed in the test banner in the CPU column.

Caveats.   None known.

Bugs.   None known.

Limitations.   None known.

Hardware Required.   Hewlett-Packard.

Software Required.   HP-UX.

Availability.   Informally supported by but not a product of Hewlett-Packard, Netperf and its documentation are nonetheless available from an area of the HP Web site, illustrated in Fig. 15-8, and found at *http://www.cup.hp.com/Netperf/NetperfPage.html.*

Figure 15-8

There's a lot to learn about Netperf.

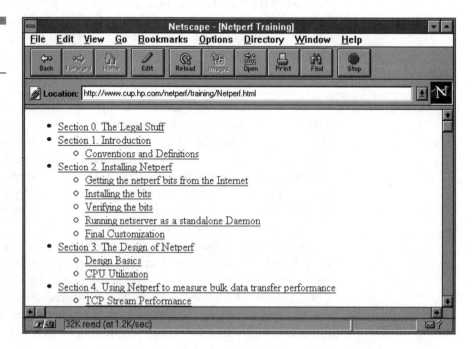

NETScout

Pertains to network status; network traffic; DECnet; IP; NFS; NMS; UNIX.

Abstract. The NETScout family of distributed LAN Analyzer devices are intended to provide network users with a comprehensive capability to identify and isolate fault conditions in data communications networks. NETScout has the capability to collect wide ranging statistical data, to display selectively captured and fully decoded network traffic, to set user-defined alarm conditions, and to obtain real-time updates from all segments of a widely dispersed internetwork from a centralized SNMP-compatible network management console.

The NETScout family is based on standards so that operation may be realized in heterogeneous networks that constitute a multiprotocol, multitopology, multivendor environment. The fundamental standards upon which NETScout is based are SNMP, which defines the protocol for all intercommunications between NETScout devices, and the Remote Monitoring Management Information Base or RMON-MIB, which defines the type of information to be gathered and made available to the user for each network segment.

NETScout clients provide a full array of monitoring and analysis features including intelligent seven-level decoding of all major protocol stacks, including TCP/IP, XNS, Novell, DECNET, LAT, APPLETALK, IBM Token Ring, Banyan Vines, and NETBIOS/SMB.

NETScout agents support all nine groups of the RMON-MIB standard. NETScout agents can work with any SNMP-based network management system and supports both Ethernet and Token Ring.

Mechanism. The operation of the NETScout family is divided into two distinct subcategories. The first is the *Client*, the user console from which operational commands are issued and where all results and diagnostic information are displayed. It is possible in a NETScout implementation to have multiple clients active simultaneously within a single network. The second category is the *Agent*, a hardware/software device attached to a specific network segment that gathers statistical information for that segment as well as providing a window into that segment where network traffic may be observed and gathered for more detailed user analysis.

Caveats. NETScout products correspond to the latest draft for Token Ring functions and will be updated as required to conform to the standard as it is approved.

Bugs. None known.

Limitations. None known.

Hardware Required. Sun SPARCstation or LattisNet Hub depending upon version.

Software Required. Sun OS 4.1.1 for client and agent, SunNet Manager for client.

Availability. NETScout is a family of commercial products. For more information regarding them, contact NetScout (formerly Frontier Software Development, Inc.) at *http://www.frontier.com.*

nfswatch

Pertains to network traffic; IP, NFS, UNIX.

Abstract. nfswatch monitors all incoming Ethernet traffic to an NFS file server and divides it into several categories. The number and percentage of packets received in each category are displayed on the screen in a continuously updated display.

By default, nfswatch monitors all packets destined for the local host over a single network interface. Options are provided to specify the specific interface to be monitored; all such interfaces may be tracked simultaneously, if desired. NFS traffic to the local host, to a remote host, from a specific host, between two hosts, or all NFS traffic on a network may be monitored.

Categories of packets monitored and counted include ND Read, ND Write, NFS Read, NFS Write, NFS Mount, Yellow Pages (NIS), RPC Authorization, Other RPC, TCP, UDP, ICMP, RIP, ARP, RARP, and Ethernet Broadcast.

Packets are also tallied either by file system or file; specific files may be watched as an option. Other packets that may be tallied include NFS RPC calls or NFS client hostname.

Facilities for taking snapshots of the tool's display screen, as well as saving data to a log file for later analysis by the included analysis tool are also available.

Mechanism. nfswatch uses the Network Interface Tap, nit(4) under SunOS 4.x, and the Packet Filter, packetfilter(4), under Ultrix 4.x, to place the Ethernet interface into promiscuous mode. It filters out NFS packets and decodes the file handles in order to determine how to count the packet.

Caveats. Because the NFS file handle is nonstandard, server-reserved data, nfswatch must be modified to understand file handles used by various operating system implementations. It currently knows about the SunOS 4.x and Ultrix file handle formats.

Bugs. None known.

Limitations. Cannot monitor FDDI interfaces. Can simultaneously monitor no more than 256 exported file systems and 256 individual files. Counts only NFS requests; NFS traffic generated by a server in response to those requests is not counted.

Hardware Required. Any VAX or DEC RISC hardware that supports Ultrix.

Software Required. Ultrix release 4.0 or later. For Ultrix 4.1, may require the patched *if_ln.o* kernel module, available from Digital's Customer Support Center.

Availability. nfswatch is copyrighted, but freely distributable, and available in several versions, as Fig. 15-9 illustrates, via anonymous FTP from *gatekeeper.dec.com*. You'll find all versions in the path */pub/net/ip/nfs*.

NOCOL

Abstract. NOCOL (Network Operations Center On-Line), one aspect of which we've depicted in Fig. 15-10, is a collection of network monitoring programs that run on UNIX systems.

Figure 15-9
One of these flavors of nfswatch may be just the tool you need.

Search results for string nfswatch

```
16/01/91   131k   /pub/net/ip/nfs/nfswatch2.0.tar.Z
23/01/91   147k   /pub/net/ip/nfs/nfswatch3.0.tar.Z
28/07/92   155k   /pub/net/ip/nfs/nfswatch3.1.tar.Z
01/03/93   252k   /pub/net/ip/nfs/nfswatch4.0.tar.Z
22/12/93   309k   /pub/net/ip/nfs/nfswatch4.1.tar.Z
```

The software consists of a number of monitoring agents that poll various parameters from any system and put it in a format suitable for postprocessing. The postprocessors can be a display agent, an automated troubleshooting program, an event logging program, and so forth. Presently, monitors for tracking reachability, SNMP traps, data throughput rate, and nameservers have been developed and are in use. A display agent NOCOL(1) using curses has already been developed.

All data collected by the monitoring agents follows a fixed, nonreadable format. Each data entry is termed an event in NOCOL, and each event has certain flags and severity associated with it. The display agent NOCOL(1) displays the output of these monitoring agents depending on the severity of the event. There can be multiple displays running simultaneously; all process the same set of monitored data.

Figure 15-10
This is the starting point for downloading NOCOL.

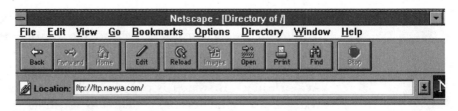

Current directory is /

```
Welcome, ftp@207.103.113.173 to the Netplex Technologies Inc. FTP Server

Files from ftp.navya.com are also stored here.

Files located under /pub are:

    NOCOL           Network Monitoring software
    XTACACS         Authentication server for Cisco Terminal Servers
```

There are four levels of severity associated with an event: CRITICAL, ERROR, WARNING, and INFO. The severity level is controlled independently by the monitoring agents, and the decision to raise or set an event's severity to any level depends on the logic embedded in the monitoring agent.

NOCOL consists of the following modules:

- *NOCOL:* displays data collected by the monitoring agents. It uses the curses screen management system to support a wide variety of terminal types. The criteria for displaying an event include:

  severity level of the event is higher than the severity level set in the display

  display filter (if set) matches some string in the event line

  The display can be in regular 80-column mode or in extended 132-column mode. Critical events are displayed in reverse video (if the terminal type supports it). Additional features, such as displaying informational messages in a part of the window, automatic resizing window sizes, operator acknowledgment via a bell when a new event goes critical, are also available.

- *ippingmon:* monitors the reachability of a site via ICMP ping packets. This program can use the default output from the system's ping program, but an accompanying program, *multiping*, can ping multiple IP sites at the same time and is preferable for monitoring a large list of sites. A site is marked unreachable if a certain number of packets is lost, and the severity level is increased each time that the site tests unreachable.

- *osipingmon:* similar to ippingmon, but uses the OSI ping program instead. The only requirement is that the system's ping program output match the typical BSD IP ping program's output.

- *nsmon:* monitors the name servers on the list of specified hosts. It periodically sends an SOA query for the default domain. If the queried name servers cannot resolve the query, the site is elevated to CRITICAL status.

- *tpmon:* monitors kilobits-per-second throughput to a list of hosts. The program connects to the discard socket on the remote machine (using a *STREAM* socket), and sends large packets for a small amount of time to evaluate the effective throughput. It elevates a site to WARNING level if the throughput drops below a certain threshold (set in the configuration file).

■ *trapmon:* converts all SNMP traps into a format suitable for displaying using NOCOL. The severity of the various traps is preset, but can be changed at compilation.

Hardware Required. Any hardware platform that supports UNIX and its *curses* management library.

Software Required. Any UNIX system with the curses screen management library and an IP programming facility. Tested under SunOS 4.1.1, Ultrix, and on NeXT systems.

Availability. Available via anonymous FTP from *ftp.navya.com.*

NPRV

Pertains to network maps; network routing; network status; IP; ping; VMS.

Abstract. NPRV is a full-screen, keypad-oriented utility that runs under VAX/VMS. It allows the user to quickly scan through a user-defined list of IP addresses (or domain names) and verify a node's reachability. The node's reachability is determined by performing an ICMP echo, UDP echo, and a TCP echo at alternating three-second intervals. The total number of packets sent and received are displayed, as well as the minimum, average, and maximum round-trip times (in milliseconds) for each type of echo. Additionally, a "trace route" function is performed to determine the path from the local system to the remote host. Once all of the trace route information has filled the screen, a "snapshot" of the screen can be written to a text file. Upon exiting the utility, these text files can be used to generate a logical network map showing host and gateway interconnectivity.

Mechanism. An ICMP echo is performed by sending ICMP ECHO REQUEST packets. The UDP and TCP echoes are performed by connecting to the UDP/TCP echo ports (port number 7). The trace route information is compiled by sending alternating ICMP ECHO REQUEST packets and UDP packets with very large destination UDP port numbers (in two passes). Each packet is initially sent with a TTL (time to live) of 1. This should cause an ICMP TIME EXCEEDED error to be generated by the first routing gateway. Then each packet is sent with a TTL of 2. This should cause an ICMP TIME EXCEEDED error to be generated by the second routing gateway. Then each packet is sent with a TTL of 3, and so on. This process continues until an ICMP ECHO REPLY or UDP PORT UNREACHABLE is received. This indicates that the remote host has been reached and that the trace route information is complete.

Caveats.   This utility sends one echo packet per second (ICMP, UDP, or TCP), as well as sending out one trace route packet per second. If a transmitted trace route packet is returned in less than one second, another trace route packet is sent in 100 milliseconds. This could cause a significant amount of contention on the local network.

Bugs.   None known. (The author, Allen Sturtevant, asks that you report any you might unearth to him at *sturtevant@ccc.nmfecc.gov.*)

Limitations.   The user is required to have SYSPRV privilege to perform the ICMP Echo and trace route functions. The utility will still run with this privilege disabled, but only the UDP Echo and TCP Echo information will be displayed. This utility is written in C, but unfortunately it cannot be easily ported over to UNIX since many VMS system calls are used and all screen I/O is done using the VMS Screen Management Routines.

Hardware Required.   Any network interface supported by TGV Incorporated's MultiNet software.

Software Required.   VAX/VMS V5.1+ and TGV Incorporated's Multi-Net version 2.0.

Availability.   For executables only, FTP to the ANONYMOUS account (password GUEST) on *CCC.NMFECC.GOV* (128.55.128.30) and GET the following files:

- NPRV.DOC (ASCII text)
- NPRV.EXE
- SAMPLE.IPA (ASCII text)

nslookup

Pertains to   network status; DNS; BIND; UNIX; VMS.

Abstract.   nslookup is an interactive program for querying DNS servers. It is essentially a user-friendly front end to the BSD BIND-resolver library routines. This program is useful for converting a host-name into an IP address, and vice versa; determining name servers for a domain; listing the contents of a domain; displaying any type of DNS record, such as MX, CNAME, and SOA; and diagnosing name server problems.

   By default, nslookup will query the default name server but you can specify a different server on the command line or from a configuration file. You can also specify different values for the options that control the resolver routines.

Mechanism.   The program formats, sends, and receives DNS queries.

Caveats.   None.

Bugs.   None known.

Limitations.   None known.

Hardware Required.   No restrictions.

Software Required.   BSD UNIX or related OS, or VMS.

Availability.   nslookup is included in the BIND distribution available with 4.xBSD UNIX and related operating systems. For VMS, available as part of TGV MultiNet IP software package. Also available as part of Wollongong's WIN/TCP.

ping

Pertains to   network status; IP; DOS, UNIX, VMS.

Abstract.   It may seem redundant to discuss ping, but it remains the most basic tool for Internet-related management: ping verifies not only that a remote IP implementation is functional, but also that intervening networks and interfaces are the same; ping can also be used to measure round-trip delay.

Mechanism.   ping is based on the ICMP ECHO_REQUEST message.

Caveats.   If run repeatedly, ping can sometimes generate high server loads.

Bugs.   None known.

Limitations.   Only one ping implementation, that included in the commercial PC/TCP suite, supports both loose and strict source routing. Though some ping implementations support the ICMP *record route* feature, the usefulness of this option for debugging is limited by the fact that many gateways do not correctly implement it.

Hardware Required.   No restrictions.

Software Required.   No restrictions.

Availability.   ping is widely included in TCP/IP distributions.

proxyd

Pertains to   network management and status; IP; NMS; UNIX.

Abstract.   SNMP proxy agents may be used to permit the monitoring and controlling of network elements that are otherwise not addressable using the SNMP management protocol (e.g., a network bridge that implements a proprietary management protocol). Similarly, SNMP proxy agents may be used to protect SNMP agents from redundant network management agents through the use of caches. Finally, SNMP proxy agents may be used to implement elaborate MIB access policies.

The proxy agent daemon

■ listens for SNMP queries and commands from logically remote network management stations

■ translates and retransmits those as appropriate network management queries or cache lookups

- listens for and parses the responses
- translates the responses into SNMP responses
- returns those as SNMP messages to the network management station that originated the transaction

The proxy agent daemon also emits SNMP traps to identified trap receivers. The proxy agent daemon is designed to make the addition of additional vendor-specific variables a straightforward task. The proxy application comes complete with source code, including a set of portable libraries for generating and parsing SNMP messages and a set of command line utilities.

Mechanism.   Network management variables are made available for inspection and/or alteration by means of SNMP.

Caveats.   None.

Bugs.   None known.

Limitations.   This application is a template for proxy application writers.

Hardware Required.   Systems from Sun Microsystems, Incorporated.

Software Required.   Sun OS 3.5 or 4.x.

Availability.   This is a commercial product available under license from SNMP Research. Visit their Web site at *http://www.snmp.com*.

query

Pertains to   routing; IP; UNIX.

Abstract.   query allows remote viewing of a gateway's routing tables.

Mechanism.   This tool formats and sends a RIP request or POLL command to a destination gateway.

Caveats.   query is intended to be used as a tool for debugging gateways, not for network management. SNMP is the preferred protocol for network management.

Bugs.   None known.

Limitations.   The polled gateway must run RIP.

Hardware Required.   No restrictions.

Software Required.   4.3BSD UNIX or related OS.

Availability.   Available with routed and gated distributions.

Sniffer

Pertains to   network traffic; DECnet; IP; NFS; SMTP.

Abstract.   Sniffer is a protocol analyzer for performing a variety of LAN diagnostics, monitoring, traffic generation, and troubleshooting, as Fig. 15-11 shows. The Sniffer protocol analyzer has the capability of capturing every packet on a network and of decoding all seven layers of the OSI protocol model. Capture frame selection is based on several differ-

Figure 15-11
Few networks would fall outside Sniffer's capabilities.

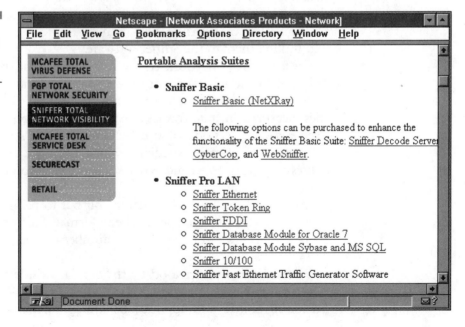

Figure 15-12
Even the most up-to-date networking technologies can be monitored by Sniffer.

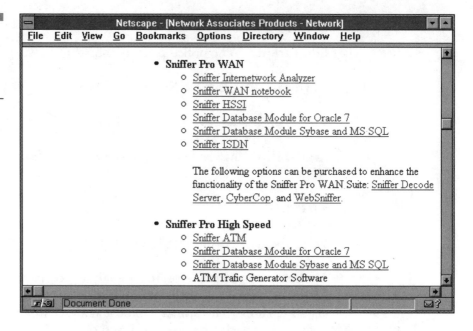

ent filters: protocol content at lower levels; node addresses; pattern matching (up to eight logically related patterns of 32 bytes each); and destination class. Users may extend the protocol interpretation capability of the Sniffer by writing their own customized protocol interpreters and linking them to the Sniffer software.

Sniffer displays network traffic information and performance statistics in real time, in user-selectable formats. Numeric station addresses are translated to symbolic names or manufacturer ID names. Network activities measured include frames accepted, Kbytes accepted, and buffer use. Each network version, some of which Fig. 15-12 shows, has additional counters for activities specific to that network. Network activity is expressed as frames/second, Kbytes/second, or percent of network bandwidth utilization.

Data collection by Sniffer may be output to printer or stored to disk in either print-file or spread-sheet format. Sniffer supports several topologies, as Fig. 15-13 shows, and a number of operating systems, as Fig. 15-14 depicts.

Protocol suites understood by the Sniffer include Banyan Vines, IBM Token-Ring, Novell Netware, XNS/MS-Net (3Com 3+), DECnet, TCP/IP (including SNMP and applications-layer protocols such as FTP, SMTP,

Figure 15-13
Token Ring, FDDI, and more are no match for Sniffer.

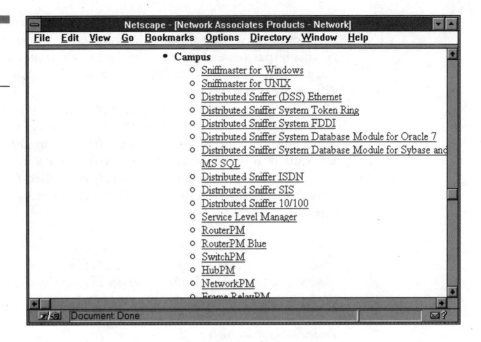

and TELNET), X Windows (for X version 11), NFS, and several SUN proprietary protocols (including mount, pmap, RPC, and YP).

Supported LANs include Ethernet, Token-Ring (4-Mb and 16-Mb versions), ARCNET, StarLAN, IBM PC Net (Broadband), and Apple Localtalk Network.

Figure 15-14
Whatever the topology, Sniffer tracks the OSes.

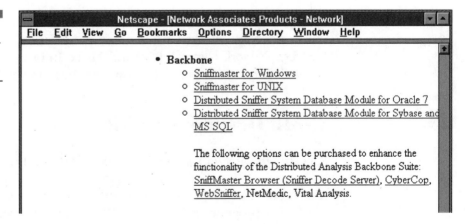

Mechanism.   Sniffer is a self-contained, portable protocol analyzer that requires only AC line power and connection to a network to operate. Normally passive (except when in Traffic Generator mode), it captures images of all or of selected frames in a working buffer, ready for immediate analysis and display.

Sniffer is a stand-alone device. Two platforms are available: one for use with single network topologies, the other for use with multinetwork topologies. Both include Sniffer core software, a modified network interface card (or multiple cards), and optional protocol interpreter suites.

All Sniffer functions may be remotely controlled from a modem-connected PC. Output from the Sniffer can be imported to database or spreadsheet packages.

Caveats.   In normal use, Sniffer is a passive device, and so will not adversely affect network performance. Performance degradation will be observed, of course, if Sniffer is set to Traffic Generator mode and connected to an active network.

Bugs.   None known.

Limitations.   None known.

Hardware Required.   None. The Sniffer is a self-contained unit, and includes its own interface card. It installs into a network as would any normal workstation.

Software Required.   None.

Availability.   Large numbers of Sniffer-related and other tools are available from Network Associates, a company formed by the merger of McAfee, Network General, PGP, and Helix. Information on these products, a little of which Fig. 15-15 shows, can be found at *http://www.nai.com*.

SNMP Development Kit

Pertains to   network status; IP; NMS; SNMP; UNIX.

Abstract.   The SNMP Development Kit contains C Language source code for a programming library that facilitates access to the manage-

Figure 15-15
Sniffing out a solution to bandwidth problems might just start here.

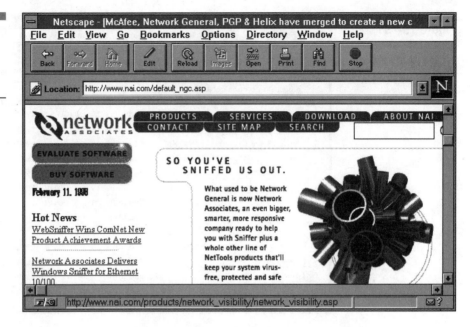

ment services of SNMP (RFC 1098). Sources are also included for a few simple client applications whose main purpose is to illustrate the use of the library. Example client applications query remote SNMP agents in a variety of modes, and generate or collect SNMP traps. Code for an example SNMP agent that supports a subset of the Internet MIB (RFC 1066) is also included.

Mechanism.   The Development Kit facilitates development of SNMP-based management applications, both clients and agents. Example applications execute SNMP management operations according to the values of command line arguments.

Caveats.   None.

Bugs.   None known. As they occur, the authors correct bugs in subsequent releases.

Limitations.   None reported.

Hardware Required.   The SNMP library source code is highly portable and runs on a wide range of platforms.

Figure 15-16
SNMP Development
Kits come in several
formats.

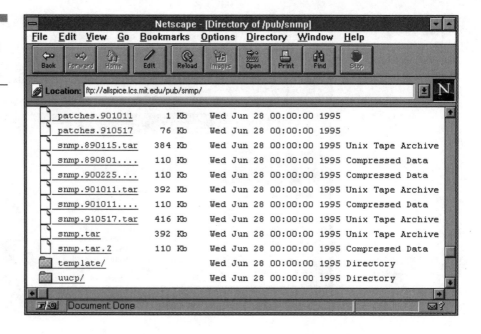

patches.901011	1 Kb	Wed Jun 28 00:00:00 1995	
patches.910517	76 Kb	Wed Jun 28 00:00:00 1995	
snmp.890115.tar	384 Kb	Wed Jun 28 00:00:00 1995	Unix Tape Archive
snmp.890801....	110 Kb	Wed Jun 28 00:00:00 1995	Compressed Data
snmp.900225....	110 Kb	Wed Jun 28 00:00:00 1995	Compressed Data
snmp.901011.tar	392 Kb	Wed Jun 28 00:00:00 1995	Unix Tape Archive
snmp.901011....	110 Kb	Wed Jun 28 00:00:00 1995	Compressed Data
snmp.910517.tar	416 Kb	Wed Jun 28 00:00:00 1995	Unix Tape Archive
snmp.tar	392 Kb	Wed Jun 28 00:00:00 1995	Unix Tape Archive
snmp.tar.Z	110 Kb	Wed Jun 28 00:00:00 1995	Compressed Data
template/		Wed Jun 28 00:00:00 1995	Directory
uucp/		Wed Jun 28 00:00:00 1995	Directory

Software Required. The SNMP library source code has almost no operating-system dependencies and runs in a wide range of environments. Certain portions of the example SNMP agent code are specific to the 4.3BSD implementation of the UNIX system for the DEC MicroVAX.

Availability. The Development Kit is, as Fig. 15-16 shows, available via anonymous FTP from *ftp.allspice.lcs.mit.edu*.

The copyright for the Development Kit is held by the Massachusetts Institute of Technology, and the Kit is distributed without charge according to the terms set forth in its code and documentation; we've shown you a bit of the latter in Fig. 15-17. The distribution takes the form of a UNIX tar file.

Requests for hard-copy documentation or copies of the distribution on magnetic media are never honored.

Simulator

Pertains to network modeling and management.

Figure 15-17
This fragment of the
Kit's documentation
discusses distribution
contents.

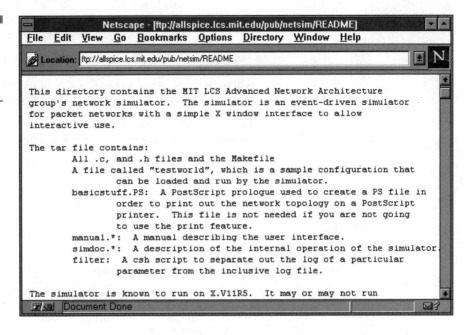

Figure 15-17
This fragment of the Kit's documentation discusses distribution contents.

Abstract. The Massachusetts Institute of Technology/ Laboratory for Computer Science Advanced Network Architecture group's network simulator is an event-driven simulator for packet networks. The simulator uses a simple X window interface to allow interactive use.

Mechanism. The simulator is distributed by means of a tar file, which contains:

- all .c, .h files, and a Makefile.

- testworld, a file containing a sample configuration that can be loaded and run by the simulator.

- basicstuff.PS, a PostScript prologue used to create a PS file in order to print out the network topology on a PostScript printer. This file is not needed if you are not going to use the print feature.

- manual.*, a manual describing the user interface.

- simdoc.*, a description of the internal operation of the simulator.

- filter, a C shell script to separate out the log of a particular parameter from the inclusive log file.

TABLE 15-2

Summarizing the
Simulator

Argument	Accomplishes
x	Don't use X windows. The simulator will not call any X routines, and so will run on a machine without X windows. Of course, no graphics output will be produced either. Although not required, it is most useful to specify a worldfile and a stoptime when using this option. Otherwise the simulator will have no world to simulate and/or will never stop. Also, the worldfile specified should be a "snapshot" that has some parameters logged to disk so that the simulator run produces some results. Not surprisingly, the simulator runs much more quickly without doing any X I/O.
s	Allows one to specify the seed for the random number generator. If this option is not given, the current time is used. The seed used is printed at the beginning of each simulator run, and is saved as a comment in any "snap" files produced by the simulator.
worldfile	A file describing the configuration of the network to simulate. Such a file is produced by the save and snap commands in the simulator.
stoptime	Length of time (in microseconds of simulated time) for the simulator to run. Most useful when running noninteractively (with the -x option). When the simulator stops, it will automatically produce a "snap" file of the current state of the world.

To make the application, one must only follow the instructions in the file INSTALL. To run it, a UNIX command line like the following must be used.

```
sim [-x] [-s seed] [worldfile [stoptime]]
```

Table 15-2 explains this generalized syntax.

Caveats. None.

Bugs. None known.

Limitations. The simulator is known to run on X.V11R5. It may or may not run under earlier versions of X11, and will almost certainly not run on R4.

Hardware Required. No restrictions.

Software Required. Any UNIX OS that supports X v11r5.

Availability. Available via anonymous FTP from *ftp.allspice.lcs.mit.edu*. The copyright for the Development Kit is held by the Massachusetts Institute of Technology, and the Kit is distributed without charge according to the terms set forth in its code and documentation. The distribution takes the form of a UNIX tar file.

Requests for hard-copy documentation or copies of the distribution on magnetic media are never honored.

SpiderAnalyzer

Pertains to network traffic; DECnet; IP.

Abstract. SpiderAnalyzer is a network analysis and management tool for Ethernet and Token Ring networks. Fully multitasking, it has all the monitoring features of the SpiderMonitor, plus additional features such as the ability to simulate traffic load in a way that closely approximates live traffic. The multisegment analyzer, used in conjunction with Spider-Probes on remote segments, provides centralized control, networkwide alarm surveillance, and remote network troubleshooting, enabling the network manager to locate and diagnose potential problems anywhere on the system.

Mechanism. SpiderAnalyzer carries out data analysis by the use of sophisticated filters and triggers, and provides detailed protocol decoding of a range of protocols spanning all seven OSI model layers.

Caveats. None.

Bugs. None known.

Limitations. Source code is not available.

Hardware Required. Sun SPARC.

Software Required. Solaris Sparc 1.0.

Availability. A commercial product available from Spider Systems Limited at Spider House, 80 Peach Street, Wokingham, RG11 1XH England. Phone: 734 771055. Fax: 734 771214.

SPIMS (Swedish Institute of Computer Science, or SICS, Protocol Implementation Measurement System)

Pertains to   IP; UNIX.

Abstract.   SPIMS is used to measure the performance of protocol and "protocol-like" services including response time (two-way delay), through-put, and the time to open and close connections. It has been used to benchmark alternative protocol implementations and to observe how performance varies when parameters in specific implementations have been varied (i.e., to tune parameters).

SPIMS currently has interfaces to the DoD Internet protocols: UDP; TCP; FTP; SunRPC; the OSI protocols from the ISODE 4.0 distribution package: FTAM, ROSE, ISO TP0, and to Sunlink 5.2 ISO TP4, as well as Stanford's VMTP. Also available are a rudimentary set of benchmarks, stubs for new protocol interfaces, and a user manual.

Mechanism.   SPIMS runs as user processes and uses a TCP connection for measurement setup. Measurements take place between processes over the measured protocol. SPIMS generates messages and transfers them via the measured protocol service according to a user-supplied specification. SPIMS has a unique measurement specification language that is used to specify a measurement session. In the language there are constructs for different application types (e.g., bulk data transfer), for specifying frequency and sequence of messages, for distribution over message sizes, and for combining basic specifications. These specifications are independent of both protocols and protocol implementations and can be used for benchmarking.

Caveats.   None.

Bugs.   None known.

Limitations.   None known.

Hardware Required.   No restrictions.

Software Required.   SPIMS is implemented on UNIX, including SunOS 4., 4.3BSD UNIX, DN (UNIX System V, with extensions), and Ultrix

2.0/3.0. It requires a TCP connection for measurement setup. No kernel modifications or any modifications to measured protocols are required.

Availability. SPIMS is not in the public domain and the software is covered by licenses. Use of the SPIMS software represents acceptance of the terms and conditions of the licenses. The licenses are enclosed in the distribution package.

Licenses and SPIMS cover letter can also be obtained via an Internet FTP connection without getting the whole software suite. The file to retrieve is *pub/spims-dist/licenses.tar.Z.*

There are two different distribution classes depending on the requesting organization: to universities and nonprofit organizations, SPIMS source code is distributed free of charge. There are two ways to get the software:

1. As Fig. 15-18 shows, by anonymous FTP to sics.se [192.16.123.90], to retrieve the file *pub/spims-dist/dist910304.tar.Z* (a .6-MB compressed tar image) in BINARY mode.

2. On a Sun $^1/_4$-inch cartridge tape. For mailing, a handling fee of US$150.00 will be charged.

Commercial organizations can chose between a license for commercial use, or a license for internal research only and no commercial use whatsoever.

Figure 15-18
Not only executables, but licensing can be found here.

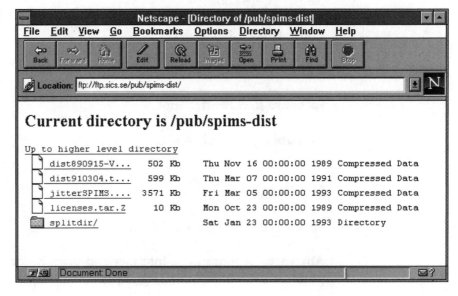

For more information about the research prototype distribution and about a commercial license, contact the Swedish Institute of Computer Science, Attn: Birgitta Klingenberg, at P.O. Box 1263, -164 28 Kista, Sweden, or by email at *spims@sics.se.*

spray

Pertains to   IP; ping; UNIX.

Abstract.   Spray is a traffic generation tool that generates RPC or UDP packets, or ICMP Echo Requests.

Mechanism.   Packets are sent to a remote procedure call application at the destination host. The count of received packets is retrieved from the remote application after a certain number of packets have been transmitted. The difference in packets received versus packets sent represents (on a LAN) the packets that the destination host had to drop due to increasing queue length. A measure of throughput relative to system speed and network load can thus be obtained.

Caveats.   Spray can congest a network.

Bugs.   None known.

Limitations.   None reported.

Hardware Required.   No restrictions.

Software Required.   SunOS.

Availability.   Supplied with SunOS.

tcpdump

Pertains to   IP; NFS; UNIX; VMS.

Abstract.   tcpdump can interpret and print headers for the following protocols: Ethernet, IP, ICMP, TCP, UDP, NFS, ND, ARP/RARP, and

AppleTalk; tcpdump has proven useful for examining and evaluating the retransmission and window management operations of TCP implementations.

Mechanism. tcpdump writes a log file of the frames traversing an Ethernet interface. Each output line includes the time a packet is received, the type of packet, and various values from its header.

Caveats. None.

Bugs. None known.

Limitations. The public domain version requires a kernel patch for SunOS. TCPware for VMS currently interprets headers for IP, TCP, UDP, and ICMP only.

Hardware Required. Any Ultrix system (VAX or DEC RISC hardware).

Software Required. Ultrix release 4.0 or later. For Ultrix 4.1, may require the patched *if_ln.o* kernel module, available from Digital's Customer Support Center.

Availability. Available, though subject to copyright restrictions, via anonymous FTP, as Fig. 15-19 depicts, from *ftp.ee.lbl.gov*, an FTP server of the Network Research Group of the Lawrence Berkeley Laboratory.

The source and documentation for the tool is in compressed tar format, in the file *tcpdump.tar.Z*. For VMS hosts with DEC Ethernet controllers, available as part of the TGV MultiNet IP software package and TCPware for VMS from Process Software Corporation.

TCPWare for VMS/SNMP Agent

Pertains to network status; network traffic; IP; VMS.

Abstract. This SNMP agent listens for and responds to network management requests sent from SNMP-conforming network management stations. The SNMP agent also sends SNMP traps, under specific conditions, to identified trap receivers. SNMP communities and generation of

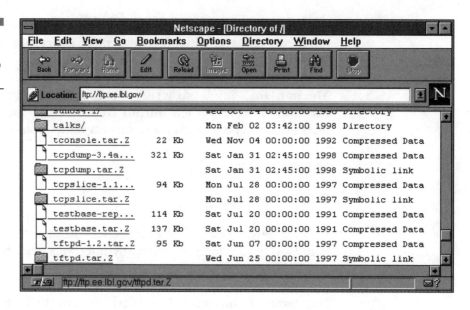

Figure 15-19
Start here to get a copy of the tcpdump executable.

traps are fully configurable. The SNMP agent supports all MIB-II variables, except the EGP group.

Mechanism. Network management variables are made available for inspection and/or alteration by means of SNMP.

Caveats. None.

Bugs. None known.

Limitations. None known.

Hardware Required. Supported VAX processors.

Software Required. VMS V4 or later.

Availability. The SNMP agent is included in TCPware for VMS, a commercial product available under license from Process Software Corporation, whose home page is shown in Fig. 15-20.

For more information on both the SNMP agent and its parent application TCPWare, visit the Process Web site at *http://www.process.com.*

Figure 15-20
As you can see,
TCPware is more
than just a collection
of monitoring tools.

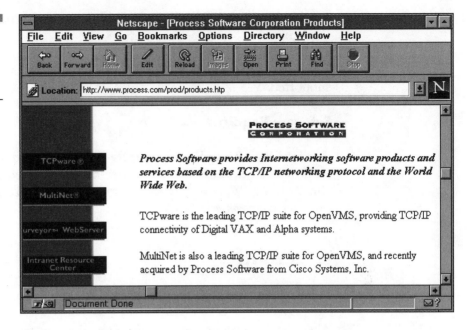

Figure 15-20
As you can see, TCPware is more than just a collection of monitoring tools.

TokenVIEW

Pertains to   network status; Token-Ring; NMS.

Abstract.   TokenVIEW is a network management tool for 4/16 Mbit IEEE 802.5 Token Ring Networks. The tool monitors active nodes and ring errors.

Mechanism.   TokenVIEW maintains a database of nodes, wire centers, and their connections. A separate network management ring used with Proteon Intelligent Wire Centers allows wire center configuration information to be read and modified from a single remote workstation. A log of network events used with a database contain nodes, wire centers and their connections, facilitates tracking and correction of network errors. Requires an "E" series PROM, sold with the package.

Caveats.   None.

Bugs.   None known.

Limitations. TokenVIEW can monitor no more than 256 nodes on a single network.

Hardware Required. 512K RAM, CGA or better, hard disk, mouse supported.

Software Required. MS-DOS.

Availability. TokenVIEW is a fully supported product of Proteon, Inc. To download TokenVIEW and other network-related Proteon tools, see the Proteon Web site, whose home page Fig. 15-21 illustrates, at *http://www.proteon.com.*

traceroute

Pertains to network traffic and routing; IP; ping.

Abstract. traceroute allows the route taken by packets from source to destination to be discovered. It can be used for situations where the IP

Figure 15-21
Proteon offers a number of network support services.

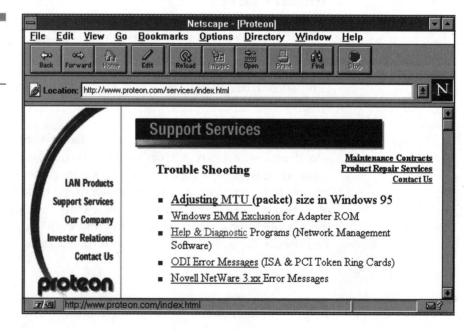

record route option would fail, such as intermediate gateways discarding packets, routes that exceed the capacity of a datagram, or intermediate IP implementations that don't support record route. Round-trip delays between the source and intermediate gateways are also reported, allowing the determination of individual gateways contribution to end-to-end delay.

Enhanced versions of traceroute have been developed that allow specification of loose source routes for datagrams. This allows one to investigate the return path from remote machines back to the local host.

Mechanism. traceroute relies on the ICMP TIME_EXCEEDED error reporting mechanism. When an IP packet is received by a gateway with a time-to-live value of 0, an ICMP packet is sent to the host that generated the packet. By sending packets to a destination with a TTL of 0, the next hop can be identified as the source of the ICMP TIME EXCEEDED message. By incrementing the TTL field, the subsequent hops can be identified. Each packet sent out is also time stamped. The time stamp is returned as part of the ICMP packet so a round-trip delay can be calculated.

Caveats. Some IP implementations forward packets with a TTL of 0, thus escaping identification. Others use the TTL field in the arriving packet as the TTL for the ICMP error reply, which delays identification.

Sending datagrams with the source route option will cause some gateways to crash. It is considered poor form to repeat this behavior.

Bugs. None known.

Limitations. Most versions of UNIX have errors in the raw IP code that require kernel mods for the standard version of traceroute to work. A version of traceroute exists that runs without kernel mods under SunOS 3.5, but it operates only over an Ethernet interface.

Hardware Required. No restrictions.

Software Required. BSD UNIX or related OS, or VMS.

Availability. A version of traceroute that supports Loose Source Record Route, along with the source code of the required kernel modifications and a Makefile for installing them, is available via anonymous FTP from *zerkalo.harvard.edu,* in directory *pub,* file *traceroute_pkg.tar.Z.*

A version of traceroute that runs under SunOS 3.5 and does *not* require kernel mods is available via anonymous FTP from *dopey.cs.unc.edu,* in the *file~ftp/pub/traceroute.tar.Z.*

For VMS, traceroute is available as part of the TGV MultiNet IP software package.

TRPT (Transliterate Protocol Trace)

Pertains to   network traffic; IP; UNIX.

Abstract.   TRPT displays a trace of a TCP socket events. When no options are supplied, TRPT prints all the trace records found in a system, grouped according to TCP connection protocol control block (PCB).

Mechanism.   TRPT interrogates the buffer of TCP trace records that is created when a TCP socket is marked for debugging.

Caveats.   Prior to using TRPT, an analyst should take steps to isolate the problem connection and find the address of its protocol control blocks.

Bugs.   None known.

Limitations.   A socket must have the debugging option set for TRPT to operate. Another problem is that the output format of TRPT is difficult.

Hardware Required.   No restrictions.

Software Required.   BSD UNIX or related OS.

Availability.   Included with BSD and SunOS distributions.

TTCP

Pertains to   IP; ping; UNIX; VMS.

Abstract.   TTCP is a traffic generator that can be used for testing end-to-end throughput. It is good for evaluating TCP/IP implementations.

Mechanism. Cooperating processes are started on two hosts. They open a TCP connection and transfer a high volume of data. Delay and throughput are calculated.

Caveats. This tool can greatly increase server load.

Bugs. None known.

Limitations. None known.

Hardware Required. No restrictions.

Software Required. BSD UNIX or related OS, or VMS.

Availability. As Fig. 15-22 shows, source for BSD UNIX is available via anonymous FTP from *sgi.com*, in the file *sgi/src/ttcp.c*. For VMS, ttcp.c is included in the MultiNet Programmer's Kit, a standard feature of the TGV MultiNet IP software package.

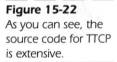

Figure 15-22
As you can see, the source code for TTCP is extensive.

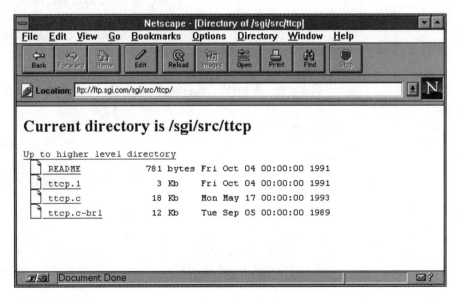

Visual Ping

Pertains to   network traffic.

Abstract.   This tool gives a graphical representation of the time taken for packets of different size to traverse the Internet between a Webserver and the client machine. This may be used in an approximation of available bandwidth. The tool cannot measure real bandwidth, or backbone bandwidth where it is greater than the bandwidth from the server.

Mechanism.   The tool consists of two scripts: a shell script, which generates an HTML document; and a Perl script, which performs the ping operations and generates an inline GIF.

For each of several different packet sizes, the target is pinged a few times. Response time is plotted against packet size. From each group, the minimum response time is taken and used in a straight line fit, which is also plotted. The fitted line is used as an indication of available bandwidth and baseline delay to the node.

The tool may run either as a CGI script on a Webserver, or stand-alone using X-11. When running stand-alone, the node to ping is given as an argument, for example, visual-ping.pl myhost.net. When running as a CGI script, the node is given as an argument using http GET. Usually the shell script visual-ping is used as a wrapper, in order to generate the GIF as an inline image. If no argument is given, the script may either use the browser's address or prompt for an address, depending on the configuration.

Caveats.   Visual Ping has been tested only on Linux and Irix.

Bugs.   None known.

Limitations.   The tool cannot measure real bandwidth, or backbone bandwidth where it is greater than the bandwidth from the server.

Hardware Required.   No restrictions.

Software Required.   The script requires Perl 4, gnuplot, ping, and optionally giftool to render the GIF transparent. It should run on any UNIX system.

Availability. Source code is available at *ftp://vancouver-webpages.com/pub/visual-ping.tar.gz.*

WebSTONE

Pertains to http; network traffic.

Abstract. The WebSTONE, a Web-serving benchmark has been developed in an attempt to better understand the performance characterics of both hardware and software platforms for HTTP. The WebSTONE tests the performance of different implementations of HTTP for a variety of server platforms. In addition, WebSTONE is also a performance tester and may be used as a tool to help identify performance characterizations of server platforms.

The WebSTONE is a configurable benchmark that can measure:

- Average and maximum connect time
- Average and maximum response time
- Data throughput rate
- Number of pages retrieved
- Number of files retrieved

The benchmark's goal is to control as much of the client environment as possible.

Mechanism. The WebSTONE is a distributed, multiprocess benchmark. The master process (WebMASTER) reads the client configuration files as well as the command line. The WebMASTER then constructs a command line for each Web child. WebMASTER then remotely spawns each child. Each of these spawned processes then reads the command line and begins communicating with WebMASTER. After all the child processes have been initialized, WebMASTER instructs the Webchildren to commence the benchmark.

As each of the children finishes its run, WebMASTER collects its data; when all child processes have submitted data, WebMASTER combines those results into a report.

WebSTONE parameters that are configurable include:

- Duration of test

- Repetition of test
- Number of files
- Number of pages
- Server software and hardware configuration
- Number of Webchildren
- Number of networks
- Number of clients
- Workload of pages
- Logging
- Debugging

What follows is part of the results of a WebSTONE test that was run for 10 minutes:

```
/usr/local/bin/webstone -w xpi0-alfalfa -c sulu:8636 -u filelist -t
10 -n %d
Client: gateweb-indy8 Number of Clients: 6
Client: gateweb-indy9 Number of Clients: 6
Client: gateweb-indy10 Number of Clients: 6
Client: gateweb-indy11 Number of Clients: 6
Waiting for READY from 24 clients
All READYs received
Sending GO to all clients
All clients started at Tue Mar 7 01:33:16 1995
Waiting for clients completion
All clients ended at Tue Mar 7 01:43:41 1995
Page # 0
Total number of times page was hit 1888
Total time 1262.655230 seconds
Maximum Response time 1.758487
Total connect time for page 21.293520
Maximum time to connect 0.038234
Total amount of data moved 23199744
Page size 12288
Total number of connects 5664
```

Caveats.   None.

Bugs.   None known.

Limitations.   None known.

Hardware Required.   No restrictions.

Software Required.   No restrictions.

Availability. WebSTONE is a product of Silicon Graphics, Inc. For more information on this tool, visit the SGI Web site at *http://www.sgi.com.*

LOOKING AHEAD

Chapter 16, the last in *Optimizing Bandwidth,* recaps the major points made in the book's first nine chapters, and then offers workarounds for each of the constraints upon bandwidth it reviews.

Workarounds

In this last chapter, we offer at least one workaround or improvement for almost every point, cited in the "In a Nutshell" sections of the book's first nine chapters, as constraining optimum bandwidth use. Altogether, this chapter presents 62 such suggestions.

Chapter 1

Constraint

Among the characteristics of TCP/IP that add bandwidth overhead are: (1) its reliance on a client/server model and (2) the practice in that model of a server's remaining in a wait state.

Workaround

1. If your server uses DHCP to supply addresses to clients, two of the three timing values involved in configuring DHCP can be set at the server. Those timing values are:

   - the renewal timer, which defines how long a client must wait while it negotiates a new lease, that is, regains the right to use, an allocated IP address

   - the rebinding timer, which defines how long a client must wait to determine if a DHCP server other than that it had originally contacted is available to service its session

2. You can also configure DHCP:

   - manually, to cause the server to do no more than inform a client of its IP address, and thereby preclude the exchange of messages that dynamic address allocation involves

   - to cause a client to store on its hard drive the address automatically assigned to it by the server, and thereby once again to eliminate some of the server/client traffic dynamic allocation generates

It's the latter form of DHCP addressing that Windows NT uses, as Fig. 16-1 illustrates.

Constraint

The inevitability of noise in any transmission.

Workaround

1. As much as possible, isolate transmission media like UTP or coaxial cables from sources of electronic interference such as

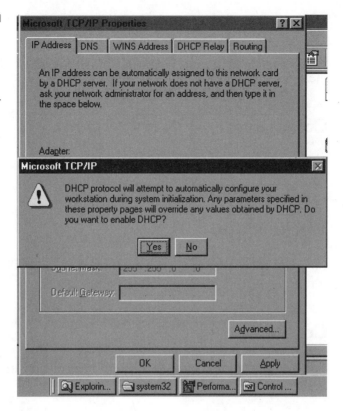

Figure 16-1
Any parameters you'd previously configured would be overridden by DHCP.

high-voltage equipment, and from sources of radio frequency interference.

2. If you use UTP, examine the wire for the number of twists per inch it contains. These twists serve to shield UTP from some forms of noise; the greater the number of twists, the more noise-free the cable is likely to be.

3. It's stating the obvious, but we'll do so anyway. Regularly check the physical condition of cables and connections. Frayed wires or dirty contacts can increase a path's likelihood of picking up noise.

Constraint

The fact that signal-to-noise ratios affect net bandwidth available.

Workaround

Consider gradually migrating to shielded twisted pair cable, nearly as inexpensive as UTP, but far less subject to noise.

Constraint

The fact that some protocols, like SLIP, not only must work with very slow serial lines but also offer no compression.

Workaround

Consider migrating to a protocol-like compressed PPP, which can offer compression while still operating across serial lines.

Constraint

The need for a variety of identifying information in a TCP/IP message header.

Workaround

Configure your TCP implementation to offer only the minimum required header, 20 octets.

Constraint

The need for a variety of routing protocols when routers are used to connect portions of heterogeneous networks.

Workaround

1. As much as possible, rely on intelligent routing protocols such as IGRP or OSPF, rather than on older, more inflexible ones like RIP.

2. If you do use OSPF, configure it to employ Variable Length Subnet Masks, which give it greater flexibility in routing packets.

Constraint

The need to establish and maintain virtual points of connection known as TCP/IP ports, so that protocols inhabiting various levels of the OSI model may communicate with one another.

Workaround

If possible, configure your TCP/IP implementation to disable such bells-and-whistles–related ports as 17 (reserved for the Quote of the Day) or 194 (set aside for the Internet Relay Char protocol).

Constraint

The need of application-level protocols like FTP to encode their instructions.

Workaround

Determine if your TCP/IP implementation, like many, carries out much of its encoding and decoding by means of APIs called by the Application Layer, rather than through the presentation layer as specified in the OSI model. If the former is the case, those APIs may be modifiable to minimize both the amount and the nature of the data encoded to be passed along to adjacent layers.

Constraint

The need of management tools like SNMP to offer a variety of parameters.

Workaround

Whether you use SNMP or some other management tool, configure it to report on only those events that are really critical to your network's efficient functioning. In similar fashion, keep the responses you configure into your management tool to the minimum needed.

Chapter 2

Constraint

The two most commonly used forms of Ethernet cabling, copper coaxial and unshielded twisted pair, can each incur bandwidth overhead due to attenuation.

Workaround

1. Repeaters can be used to boost signal on either of these media.
2. A variety of repeater called a *multiport repeater* not only regenerates signals but also can link segments.

Constraint

In addition to attenuation, coax can encounter bandwidth overhead consisting of propagation delay engendered by repeaters.

Workaround

1. Should your network experience propagation delay on segments serviced by repeaters, the strong possibility exists that those repeaters contribute to that delay. Consider, as a solution:
   - reducing the number of repeaters you use on a given segment, and throughout your environment
   - substituting a bridge for any repeater that serves in large part to connect segments
2. Since thick Ethernet cable transfers or propagates a signal at a slightly faster rate than thin Ethernet, take the nature of your cabling system into account when positioning repeaters.

Constraint

Although 10 Base T networks, because they cannot be daisy-chained, don't experience repeater-generated propagation delay, they can present other forms of bandwidth overhead: those due to exceeding the 5-4-3 rule and thereby making collisions more likely.

Workaround

We can only state the obvious: never break this rule.

Constraint

Configuring individual segments to a length that precludes collision messages being placed on the network in a timely fashion.

Workaround

Where possible, reconfigure your network to minimize the length of 10 Base T cable runs, keeping them as far below the maximum of 100 meters as practical.

Chapter 3

Constraint

The bus topology can create bandwidth overhead through its handling of collisions.

Workaround

Consider moving, if possible, to a TCP/IP implementation whose algorithms for retransmission and timeout intervals are more flexible and adaptive. The first of these, for instance, controls the amount of time a station must wait to retransmsit after having been involved in a collision.

Constraint

The ring topology, although ordinarily performing well, can experience delays under high-traffic conditions.

Workaround

If your environment regularly experiences delays of this sort, consider:

- limiting the applications available to stations that contribute most heavily to such delays
- staggering the times during which stations are available

Constraint

A number of hybrid topologies such as the star-wired ring seek to blend the best characteristics of their source topologies, while avoiding those sources' shortcomings.

Workaround

Be careful, in implementing a hybrid topology, that you adhere to the following commonsense rules for mixed networks.

- Avoid maximums within your network, including maximum lengths for cable runs in given segments, and maximum numbers of segments or repeaters.
- Be sure to adhere to the 5-4-3 rule.
- Use repeaters to connect segments of different topologies or media.

Constraint

LAN-to-WAN connections offer their own collection of bandwidth problems, since the interface between the two types of networks can operate no faster than the WAN itself.

Workaround

1. If, as many are, your LAN-to-WAN connection is a dial-up one, use the fastest modems available.
2. Configure your dial-up remote access services to minimize such factors as length of time a remote session may remain inactive before being disconnected.

Chapter 4

Constraint

Standard UTP-based dial-up connections cannot transfer at rates faster than 56 Kbps.

Workaround

Since this is the case and will continue to be so for at least the immediate future, it becomes that much more important to configure your dial-up remote access to avoid bandwidth wasted by circumstances like lengthy inactive sessions.

Constraint

Dial-up connections are inadequate for the transfer of multimedia material.

Workaround

Severely limit, or even deny, the access to multimedia-based applications that can be gained by dial-up connections.

Chapter 5

Constraint

The synchronization required by sending and receiving modems adds bandwidth overhead to any environment.

Workaround

Use the fastest modems available. They'll still have to synchronize, but will make up for the process more quickly.

Constraint

The software modules used by NIC packet drivers, although they do not directly draw upon bandwidth, do affect the speeds at which data communications take place within LANs.

Workaround

Here again, your only real recourse is to make sure you've got the fastest card available.

Constraint

The number of MAC addresses bridges can store affects their use of bandwidth by influencing the speed at which they can forward packets.

Workaround

In deciding upon a bridge, you must also choose between two categories: transparent and source-routing. These two types affect bandwidth in different ways.

1. *Transparent bridges* are those that maintain tables of MAC addresses as described in Chap. 6. If a packet arrives at such a bridge for which the bridge has no match in its address table of (destination) addresses, the bridge will forward that packet to all addresses of which it's aware (except, of course, the one from which the packet arrived). Such situations can be caused by incorrectly configured bridges. To avoid these scenarios, make sure you place transparent bridges optimally within your network. In particular:

   - avoid physical loops, that is, placements of bridges that might contribute to a message's circulating among them

   - configure your bridges to be aware of only those MAC addresses they really need to know about

2. In dealing with source-routing bridges, you're in a different ball game. *Source-routing bridges* function much like routers, at least in the sense that they attempt to determine the best path to a destination. Source-routing bridges must store such information; they limit the number of hops a message can take to 13, since they can hold only 14 routing-related fields in their internal tables. So, if you're dealing with a source-routing bridge:

   ■ place it physically within your network in such a way as to avoid generating longer-than-needed prospective routes

   ■ configure it to make the routing information it maintains as terse as possible

Constraint

In considering the role of routers and hubs in bandwidth consumption, not only the size of these devices' address tables but also such characteristics as their packet filtering and data forwarding rates must be taken into account.

Workaround

1. Look for the best balance among all these characteristics. For instance, if yours is an environment in which a large address table, but little filtering, is needed, forgo the latter capability in deciding upon a router or hub.

2. The factor that most significantly affects router transfer rates is the routing protocol it uses. Routers that rely on flexible, intelligent routing protocols like OSPF and IGRP are preferable to those that use less adaptable routing protocols such as RIP.

Chapter 6

Constraint

The use of arp and rarp to translate between IP and physical addresses can add to a network's bandwidth requirements.

Workaround

1. Use the arp command, available under both UNIX and Windows NT, to monitor arp activity on your network. Under UNIX, you can configure the tables used by the arp command flexibly, as Fig. 16-2 shows.

2. Under UNIX, the arp command not only monitors but allows you to modify the address translation tables of this protocol. Use this modification ability to remove redundant or extinct entries.

Constraint

Operating systems like UNIX and Windows NT offer means of monitoring a server's arp cache, and thereby of fine-tuning one aspect of bandwidth performance.

Figure 16-2

Here's a little of what UNIX on-line documentation tells us about running the arp command under HP-UX 10.2.

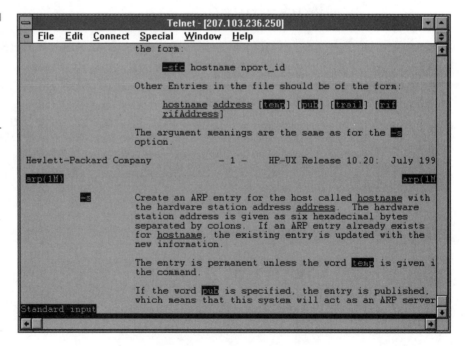

Workaround

As we've seen, UNIX provides flexibility in configuring the arp command itself, and through it the way in which the protocol it monitors will function. Windows NT offers a similar, but not quite as extensive, capability. Under NT, the arp command lets you:

- display the contents of the arp cache
- remove entries from the arp table
- add entries to that table

However, you can't, under NT, make entries temporary or permanent, as you can in UNIX. So, when using NT's arp protocol watchdog, you must exercise more caution. For instance, don't add entries to the arp table that represent hosts that are not yet up and running on your network.

Be careful too in using NT's arp to supply appropriate addresses. As Fig. 16-3 demonstrates, failing to supply an address might result in edits to the arp table of a device you hadn't meant to reconfigure.

Constraint

Static routing offers less opportunity for incurring bandwidth overhead than does dynamic routing, since the latter relies on additional protocols.

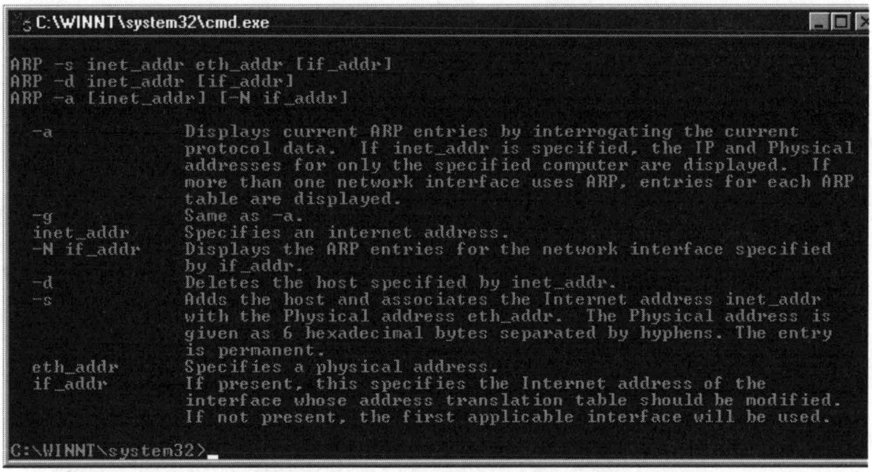

Figure 16-3
Read this excerpt from NT's on-line documentation and consider how a missing address might affect arp tables.

```
C:\WINNT\system32\cmd.exe                                         _ □ ×

ARP -s inet_addr eth_addr [if_addr]
ARP -d inet_addr [if_addr]
ARP -a [inet_addr] [-N if_addr]

     -a           Displays current ARP entries by interrogating the current
                  protocol data.  If inet_addr is specified, the IP and Physical
                  addresses for only the specified computer are displayed.  If
                  more than one network interface uses ARP, entries for each ARP
                  table are displayed.
     -g           Same as -a.
     inet_addr    Specifies an internet address.
     -N if_addr   Displays the ARP entries for the network interface specified
                  by if_addr.
     -d           Deletes the host specified by inet_addr.
     -s           Adds the host and associates the Internet address inet_addr
                  with the Physical address eth_addr.  The Physical address is
                  given as 6 hexadecimal bytes separated by hyphens. The entry
                  is permanent.
     eth_addr     Specifies a physical address.
     if_addr      If present, this specifies the Internet address of the
                  interface whose address translation table should be modified.
                  If not present, the first applicable interface will be used.

C:\WINNT\system32>
```

Workaround

Depending upon the configuration of your network, you can combine static routing with dynamic in an effort to achieve an optimal combination of ease of maintenance of routing tables and minimization of overhead attributable to routing protocols. For example, if a network contains a segment whose physical layout and services offered will not change, it might be feasible to connect that segment to the others in the net by static routing.

Constraint

RIP, a common dynamic routing protocol, can contribute to bandwidth overhead through its practices of advertising at regular intervals, and of discarding messages and even routes after certain periods of time.

Workaround

1. Many implementations of RIP allow you to configure the interval at which routers running this protocol will advertise. If yours is one of these, configure that interval to a value, determined by monitoring traffic on the segments served by such routers, which will provide sufficient but not superfluous advertising.
2. Routers running RIP can also be configured as passive routers, that is, devices that do not advertise at all but rather receive only routing information. Where it's appropriate, such configuration can reduce bandwidth overhead.

Constraint

Utilities like traceroute can be valuable tools in monitoring performance of a network as a whole, and of individual network segments, routes, and routers.

Workaround

As Figs. 16-4 through 16-6 show, even the seemingly rudimentary utility ping can supply routing diagnostics. For instance, the -o option of ping

Figure 16-4

You can configure ping in a number of ways.

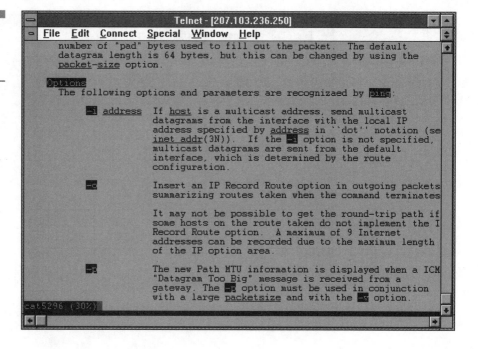

```
Telnet - [207.103.236.250]
File  Edit  Connect  Special  Window  Help
number of "pad" bytes used to fill out the packet.  The default
datagram length is 64 bytes, but this can be changed by using the
packet-size option.

Options
    The following options and parameters are recognizaed by ping:

    -i address    If host is a multicast address, send multicast
                  datagrams from the interface with the local IP
                  address specified by address in ``dot'' notation (se
                  inet addr(3N)).  If the -i option is not specified,
                  multicast datagrams are sent from the default
                  interface, which is determined by the route
                  configuration.

    -R            Insert an IP Record Route option in outgoing packets
                  summarizing routes taken when the command terminates

                  It may not be possible to get the round-trip path if
                  some hosts on the route taken do not implement the I
                  Record Route option.  A maximum of 9 Internet
                  addresses can be recorded due to the maximum length
                  of the IP option area.

    -P            The new Path MTU information is displayed when a ICM
                  "Datagram Too Big" message is received from a
                  gateway. The -P option must be used in conjunction
                  with a large packetsize and with the -v option.
cat5296 (30%)
```

Figure 16-5

Using the -v option of ping allowed us to see the transit times of the packets it sent.

```
Telnet - [207.103.236.250]
File  Edit  Connect  Special  Window  Help

PING 207.103.0.2: 64 byte packets
64 bytes from 207.103.0.2: icmp_seq=0. time=10. ms
64 bytes from 207.103.0.2: icmp_seq=1. time=9. ms
64 bytes from 207.103.0.2: icmp_seq=2. time=10. ms
64 bytes from 207.103.0.2: icmp_seq=3. time=9. ms
64 bytes from 207.103.0.2: icmp_seq=4. time=9. ms
64 bytes from 207.103.0.2: icmp_seq=5. time=10. ms
64 bytes from 207.103.0.2: icmp_seq=6. time=30. ms
64 bytes from 207.103.0.2: icmp_seq=7. time=18. ms
64 bytes from 207.103.0.2: icmp_seq=8. time=28. ms
64 bytes from 207.103.0.2: icmp_seq=9. time=10. ms
64 bytes from 207.103.0.2: icmp_seq=10. time=19. ms
64 bytes from 207.103.0.2: icmp_seq=11. time=88. ms
64 bytes from 207.103.0.2: icmp_seq=12. time=16. ms
64 bytes from 207.103.0.2: icmp_seq=13. time=12. ms
64 bytes from 207.103.0.2: icmp_seq=14. time=10. ms
64 bytes from 207.103.0.2: icmp_seq=15. time=10. ms
64 bytes from 207.103.0.2: icmp_seq=16. time=9. ms
64 bytes from 207.103.0.2: icmp_seq=17. time=9. ms
64 bytes from 207.103.0.2: icmp_seq=18. time=10. ms
64 bytes from 207.103.0.2: icmp_seq=19. time=11. ms
64 bytes from 207.103.0.2: icmp_seq=20. time=10. ms
64 bytes from 207.103.0.2: icmp_seq=21. time=102. ms
64 bytes from 207.103.0.2: icmp_seq=22. time=11. ms
#
```

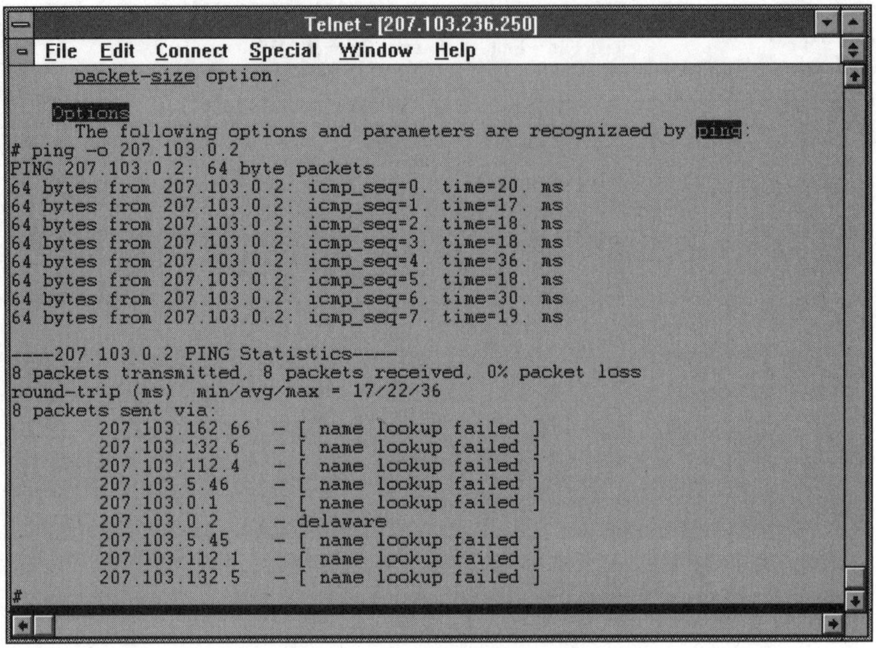

Figure 16-6

Ping's -o option produced this report.

```
 ▬ ▬ ▬ ▬                                    Telnet - [207.103.236.250]              ▼ ▲
  ⊟  File   Edit   Connect   Special   Window   Help                                  ♦
          packet-size option.                                                        ♦

      Options
          The following options and parameters are recognizaed by ping:
  # ping -o 207.103.0.2
  PING 207.103.0.2: 64 byte packets
  64 bytes from 207.103.0.2: icmp_seq=0. time=20. ms
  64 bytes from 207.103.0.2: icmp_seq=1. time=17. ms
  64 bytes from 207.103.0.2: icmp_seq=2. time=18. ms
  64 bytes from 207.103.0.2: icmp_seq=3. time=18. ms
  64 bytes from 207.103.0.2: icmp_seq=4. time=36. ms
  64 bytes from 207.103.0.2: icmp_seq=5. time=18. ms
  64 bytes from 207.103.0.2: icmp_seq=6. time=30. ms
  64 bytes from 207.103.0.2: icmp_seq=7. time=19. ms

  -----207.103.0.2 PING Statistics-----
  8 packets transmitted, 8 packets received, 0% packet loss
  round-trip (ms)  min/avg/max = 17/22/36
  8 packets sent via:
          207.103.162.66   - [ name lookup failed ]
          207.103.132.6    - [ name lookup failed ]
          207.103.112.4    - [ name lookup failed ]
          207.103.5.46     - [ name lookup failed ]
          207.103.0.1      - [ name lookup failed ]
          207.103.0.2      - delaware
          207.103.5.45     - [ name lookup failed ]
          207.103.112.1    - [ name lookup failed ]
          207.103.132.5    - [ name lookup failed ]
  #
```

under HP-UX 10.2 allows you to tell ping to inform you about routes taken by its test packets.

Or, you can cause ping to do more than echo those packets. In any case, ping can go beyond telling you if a destination is available.

Other utilities such as netstat, shown in Fig. 16-7, also provide valuable performance information.

Constraint

In environments that must ensure backward protocol compatibility as well as future protocol expansion, technologies like SNAP can be employed.

Workaround

SNAP adds its own header information to an IP header; that added information includes definitions of the protocol for which a message

Figure 16-7
Netstat can be executed in a variety of ways, to provide a variety of performance data.

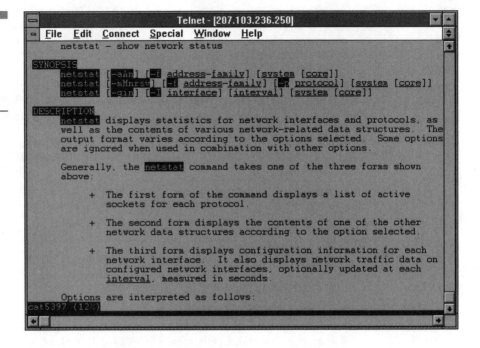

may be intended, as well as an indication of the data frames' type. So, SNAP increases the amount of information that must be handled when an IP header is processed. However, because of the additional information it provides, SNAP allows communication between:

- 802.3 Ethernet, whose frame structure as defined by the IEEE doesn't present information about the protocol for which a packet is destined, and Ethernet-II, whose frame structure, still that of the original DIX Ethernet specifications, does offer such information

- AppleTalk networks or segments and TCP/IP ones

- Token-Ring networks or segments and TCP/IP ones

Constraint

In order to aid efficient delivery of streamed information, protocols like VDP must be evolved.

Workaround

In the meantime, you can make the delivery of streamed information as efficient as possible by:

- limiting the number of applications that offer such data by, for example, forgoing video if it doesn't really add to the information content of a distributed application

- using appropriate compression techniques to compact streamed data

Chapter 7

Constraint

The networking subsystems of many widely used operating systems, including popular flavors of UNIX, are, like the OS as a whole, interrupt-driven.

Workaround

1. Configure distributed applications' execution priorities in such a way as to minimize the effect on protocol processing of an application's ability to interrupt that processing.

2. Regularly monitor the usage levels of distributed applications, and reconfigure, for instance for limited access, or even remove any that are lightly used.

3. Configure services like HTTP, or tools like firewalls, to minimize their impact on protocol processing.

Constraint

The standard networking subsystem of BSD UNIX signals the arrival of *every packet* with an interrupt.

Workaround

Consider modifying your networking subsystem to use asynchronous packet transmission, thereby avoiding the overhead needed to acknowledge every packet generated.

Constraint

When transmitting, the standard networking subsystem of BSD UNIX queues packets to a network interface, whose interrupt handler in turn controls actual transmission of those packets.

Workaround

Assign appropriate but high execution priorities to NIC device drivers.

Constraint

BSD's standard networking subsystem charges CPU time required by network packet processing, not to the interrupts or handlers that requested that processing, but rather to the user applications that generated the network request, thereby possibly unfairly affecting subsequent allocation of the CPU to these applications.

Workaround

Once again, take commonsense steps like configuring distributed applications' execution priorities to minimize the effect on protocol processing, and reconfiguring lightly used applications.

Constraint

When running under UNIX, HTTP servers are particularly prone to hanging because of circumstances like broadcast and multicast traffic.

Workaround

Use broadcast and multicast sparingly on your Web site, including such capabilities only when they truly add to information content.

Constraint

Incorrect protocol configuration at the client level.

Workaround

Fine-tune clients' http configuration to accomplish things like:

- limiting the areas of a Web site's content to which the client has access
- limiting the number of retrieval requests a given client can make
- limiting the length of time a client session can remain inactive

Constraint

Simultaneous requests from a very large number of clients to establish a TCP connection.

Workaround

Tailor the number of simultaneous http requests your server will satisfy.

Chapter 8

Constraint

Windows NT Server's combination of support for multiple protocols.

Workaround

Don't configure NT to offer protocols you may not need.

Constraint

Use of the NetBIOS application interface.

Workaround

If you configure NT Server to use the Windows Internet Naming Service or WINS, you can avoid the need, which would otherwise be present, to issue requests for address resolutions of this sort by means of IP broadcast messages. With WINS, address resolution requests are made directly to the service's database.

Constraint

Employing the Microsoft Transport Driver Interface (TDI) to map protocol addressing to NetBIOS.

Workaround

TDI is intended to provide a set of rules by which upper-layer services such as applications can communicate not only with transport protocols but also with the NDIS interface that allows any of the protocols NT supports to run over a single adapter. It follows, therefore, that limiting the number of such protocols allows TDI to work as efficiently as possible.

Constraint

The Windows NT TCP/IP driver uses a NetBIOS compatibility layer over TCP/IP itself.

Workaround

1. Once again, employing WINS streamlines NetBIOS/TCP conversations by making the address resolution involved more efficient.

2. Use the Windows NT tool nbtstat to monitor the NetBIOS/TCP/IP name/address relationship.

3. Use the NT tool netstat, which we've illustrated in Fig. 16-8 and which is very much analogous to the UNIX command of the same name, to monitor the status of such NetBIOS/TCP pairings.

Constraint

Microsoft recommends that Windows NT servers rely on the dynamic host configuration protocol to assign IP addresses, despite the overhead inherent in using DHCP.

Workaround

If you configure DHCP to operate automatically rather than dynamically, that is, to assign a permanent address to a client instead of shifting a client among available addresses, much of this overhead can be precluded.

Figure 16-8
Netstat reports on the status of NetBIOS/TCP pairs.

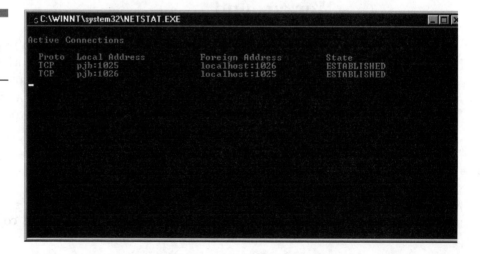

Constraint

Windows NT Server, because it can maintain only static routing tables, must rely on third-party routers for dynamic routing.

Workaround

Make sure that the router you choose relies on a flexible, intelligent routing protocol like OSPF or IGRP.

Constraint

Can thereby incur the bandwidth overhead typical of such routers' broadcasts to one another.

Workaround

1. Configure your router's broadcast intervals to provide a balance between ensuring accurate network route status information and minimizing the bandwidth consumed by routers' informational broadcasts.
2. If appropriate, make some routers passive, thereby preventing them from contributing to bandwidth consumption generated by routers' informational broadcasts.

Constraint

Windows 95, when relying on the point-to-point tunneling protocol, may experience bandwidth overhead caused by that protocol's sometimes doubling up routable and poorly or not-routable protocols, for example, carrying NetBEUI messages on the back of TCP/IP packets.

Workaround

Use the built-in NT command tracer to monitor the effect of such piggybacking on packets' travel through a network.

Chapter 9

Constraint

Apache's performance can be adjusted by configuring it at compilation to include or exclude performance-related modules, and by setting certain of its configuration directives to optimize performance.

Workaround

Be sure you really need a module before you compile it in.

Constraint

Among the Apache modules most likely to affect throughput are: mod_access, mod_alias, mod_auth, mod_auth_anon, mod_cern_meta, mod_imap, mod_include, mod_mime, mod_negotiation, mod_proxy, and mod_rewrite.

Workaround

If you don't need 'em, don't compile 'em.

Constraint

Among the configuration directives whose settings can affect Apache's resource and bandwidth consumption are MaxClients, HostnameLookups, and FollowSymLinks.

Workaround

Appendix B offers a complete list of the configuration directives Apache makes available.

Constraint

By introducing additional protocols and the need for additional inter-protocol conversations, Internet Information Server also introduces the potential—in its ADSI, IAS, IMS, and INS modules—for bandwidth overhead.

Workaround

As we've mentioned several times throughout the book and particularly in this chapter, don't configure options you don't really need. Resisting the temptation to include bells and whistles is particularly important when configuring Internet Information Server; it offers so many. But if your users don't really need streamed video, for example, forgo netShow and its bandwidth appetite.

LOOKING AHEAD

The remainder of this book offers a number of sources of additional information on bandwidth-related topics.

APPENDIX A

Introduced only about 20 years ago, the TCP/IP protocol suite has come to dominate the Internet. These protocols are not only the most widely implemented in the industry today, but also are available from virtually every computer vendor. This appendix offers a brief orientation to the most important members of the TCP/IP suite.

TCP

TCP is a connection-oriented transport protocol that sends data as an unstructured stream of bytes. By using sequence numbers and acknowledgment messages, TCP can provide a sending node with delivery information about the packets it transmits. Should data be lost in transit from source to destination, TCP can retransmit that data; only successful delivery or a timeout causes it to stop resending.

TCP recognizes duplicate messages and discards them. If the sending computer transmits too fast for the receiver, TCP can bring flow-control mechanisms to bear in order to slow data transfer. TCP can also disseminate delivery information to the upper-layer protocols and applications it supports.

IP

IP is the most important routing protocol in the TCP/IP family. IP provides not only internetwork routing, but also:

- error reporting
- disassembly and reassembly of messages, in order to be able to move them between networks with different maximum data unit sizes

IP Addresses

IP addresses are unique, 32-bit numbers that allow networks anywhere in the world to communicate with each other. An IP address is divided into three parts:

- network address
- subnet address
- host address

IP addressing supports three different network classes:

- Class A
- Class B
- Class C

Table A-1 summarizes these IP network classes. In all three cases, the left-most bit or bits indicate the network class.

Subnets

IP networks also can be divided into smaller units called *subnetworks* or *subnets*. For example, assume that a network has been assigned a Class A

TABLE A-1

Outlining Network Classes

These Networks	Are	and Use
Class A	Made up of as many as 16 million machines	8 bits in the network address field
Class B	Moderate-sized networks	16 bits in the network address field
Class C	Small networks	24 bits in the network address field
Class D	Usually reserved for multi-casting	

address and all the nodes on the network use a Class A address. Assume too that the dotted decimal representation of this network's address is 34.0.0.0. Syntax of this sort, which places only zeroes in the portions of the IP address that represent the network's host, indicate an entire network. To subdivide such a network using subnetting, an administrator must use bits that would otherwise have been available to the host portion of an IP address, specifying them as a subnet field.

If, for instance, an administrator uses 8 bits of his or her IP network's address for subnetting, the second octet of the address defines the subnet number. In the example in the last paragraph, address 34.1.0.0 therefore indicates network 34, subnet 1.

The number of bits that a subnet address can borrow in this way varies. IP uses the concept of a subnet mask to specify how many bits a subnet address uses, and where those bits fall in the host field. Subnet masks follow the same format as IP addresses, and place zeroes in all bits except those that specify the host field. So, a subnet mask that sets aside 8 bits of a Class A address to define subnetting would be written as 255.255.0.0, whereas a mask that specifies 16 bits of subnetting for Class A addresses would be given as 255.255.255.0.

Subnet masks can be passed through a network on demand so that new nodes can learn how many bits of subnetting are being used on their network. All subnets of the same network ordinarily use the same subnet mask. However, particularly in heterogeneous networks, where, for example, some subnets have several hosts and others only a few, this uniformity of subnet masks can waste address space, since each subnet consumes an entire subnet number.

As IP subnets have grown, administrators have looked for ways to use their address space more efficiently. One such technique is that of *variable length subnet masks* or VLSM. Under VLSM, a network administrator can use a long mask on networks with few hosts and a short mask on subnets with many hosts. However, as you might imagine, VLSM is more difficult to configure than ordinary subnet masking. What's more, in order to use VLSM, a network administrator must use a routing protocol that supports it. Among such protocols are:

- Open Shortest Path First (OSPF)
- Intermediate-System-to-Intermediate-System (IS-IS)
- Enhanced Interior Gateway Routing Protocol (Enhanced IGRP)
- static routing

Dynamic Addressing

On some media, such as Ethernet, IP addresses can be assigned or, in the parlance, *discovered* on the fly, through either of two other members of the TCP/IP suite:

- Address Resolution Protocol (ARP)
- Reverse Address Resolution Protocol (RARP)

Table A-2 summarizes these addressing protocols.

Routing in IP Environments

Routers, whether they reside on an Intranet or serve some section of the Internet, are organized hierarchically. Some routers are used to move information through autonomous systems, that is, across one particular group of networks under the same administrative authority and control. Such routers, which exchange information within autonomous systems, are known as *interior routers,* and use a variety of *interior gateway protocols* or IGPs to do their jobs. On the other hand, routers that move information between autonomous systems are called *exterior routers,* and use one of two exterior routing protocols:

- *exterior gateway protocol* or EGP
- *border gateway protocol* or BGP

Whether interior or exterior, TCP/IP routing protocols are dynamic, and therefore require that routers calculate routes. The algorithms used

TABLE A-2	**This Protocol**	**Determines Addresses by**
Outlining ARP and RARP	ARP	Using broadcast message to find out the hardware, that is, the MAC layer address corresponding to a particular network-layer address. ARP is generic enough to allow IP to work with any type of media access.
	RARP	Using broadcast messages to determine the network-layer address associated with a particular hardware address. RARP works primarily with diskless nodes, whose network-layer addresses often are unknown at boot time.

to do so, if they are to function efficiently, must:

- adapt to changes in a network
- automatically select optimum routes

Routing Tables.   IP routing tables are simple files, made up of only pairs of addresses. Each such pair defines a single instance of a message's destination, and the next hop it must take to reach that destination.

IP routing requires that messages hop only one router at a time. For this reason, a message never knows its complete route when it starts its journey. Rather, at each stop—that is, at every router it traverses—the message's next hop is defined by matching the destination address within the message to an entry in the current router's routing table. Therefore, any single router does no more than forward packets based on such internal information.

ICMP.   IP notify a message's source of routing problems. Another TCP/IP protocol, the Internet Control Message Protocol or ICMP, handles this job.

In addition to its most important reason for being, ICMP can carry out several other sorts of tasks in an IP network. ICMP can offer:

- a means of testing node reachability across a network, through its Echo and Reply messages
- a way to inform sources that a message has exceeded its allocated time to exist in transit, through the Time Exceeded message
- the ability to improve routing efficiency, by means of its Redirect message

Interior Routing Protocols.   In this section, we briefly examine several IGPs that are widely used by TCP/IP networks.

RIP.   Any investigation of TCP/IP routing protocols should begin with a review of the granddaddy of them all, the *routing information protocol* or RIP. RIP was developed by Xerox Corporation in the early 1980s; today, many networks still employ some form of RIP.

Although efficient in small, homogeneous networks, RIP has significant limitations in the context of larger nets, among which are the following:

- RIP determines paths according to only one criterion: the number of hops between end nodes, and ignores many factors relevant to efficient routing, such as differences in line speed.

- RIP is slow to converge; that is, it uses a relatively long time to inform all routers of changes in network status.

- RIP limits to 16 the number of router hops between any two hosts in a network.

IGRP. The *interior gateway routing protocol* (IGRP), devised by Cisco Systems in the early 1980s but now widely used, takes a tack different from that of RIP. IGRP determines best path by examining bandwidth and delay on segments or networks between routers. In addition, IGRP converges faster than RIP and lacks the earlier routing protocol's hop-count limit, which as a result, avoids routing loops such as those caused by disagreement over next hops.

NOTE: Cisco Systems recently introduced an enhanced version of IGRP called, logically enough, Enhanced IGRP. Enhanced IGRP offers both distance-vector and link-state routing algorithms. See Chap. 6 for a more detailed presentation of routing algorithms.

Enhanced IGRP has a number of attractive features, including:

- its ability to handle routing for non-TCP/IP networks, thereby providing route management for such NOSes as AppleTalk and Novell

- its significantly smaller appetite for bandwidth, which results from its ability to limit the exchange of routing information to changed information only

OSPF. Another interior gateway protocol intended as an improvement upon RIP is the *open shortest path first* or OSPF protocol. OSPF was developed by the Internet Engineering Task Force and is supported by every significant manufacturer of IP routers.

OSPF is an intradomain, link-state, hierarchical routing protocol, which, in a nutshell, means that OSPF allows network administrators to divide autonomous systems into routing areas. These areas most frequently are made up of one or more logically and physically related subnets.

Although it presents a potential drawback in requiring all routing areas to connect to a network's backbone, OSPF has the advantages of offering:

- fast rerouting
- variable length subnet masks

IS-IS. The *integrated-system-to-integrated-system* or IS-IS routing protocol, defined in the ISO 10589 specification, is, like OSPF, an intradomain, link-state, hierarchical routing protocol. IS-IS shares other characteristics with OSPF, such as the ability to operate over a variety of subnetworks, including broadcast LANs, WANs, and point-to-point links.

Exterior Routing Protocols. EGPs provide routing between, rather than within, autonomous systems. The two most popular EGPs in the TCP/IP community are discussed in this section.

EGP. The first exterior routing protocol to gain wide acceptance and use was the *exterior gateway protocol* or EGP. EGP:

- uses dynamic routing
- assumes that all autonomous systems are members of a tree topology

EGP uses no metrics, and therefore cannot make fully intelligent routing decisions. Rather, EGP routing updates consist strictly of network reachability information.

BGP. Like EGP, the *border gateway protocol* or BGP was created for use in Internet core routers. But unlike EGP, BGP was designed to help prevent routing loops and to allow more intelligent route selection. In its most recent incarnation, BGP4, this protocol also addresses scaling problems related to the explosive growth of the Internet.

Access Restrictions

Routers of whatever manufacture most frequently employ what are called *access lists* to manage network access security. Such lists are a router's basis for preventing packets with specific characteristics from entering or leaving a network. More specifically, an access list is a sequential list of instructions that, when executed, either permit or deny access through a router. Access lists can, for example:

- allow access to a LAN only to messages being presented by means of specific protocols such as FTP or HTTP

- deny access to all computers on one network segment but permit access from all other segments

- permit TCP connections from a local segment to any host in the Internet, but deny connections from the Internet to any local host

In addition to the security measures offered by access lists, access restrictions (especially those specified by the Department of Defense in RFC 1108 of the IP Security Option or IPSO standard), can be accomplished through security extensions to IP. There are two such extensions, the Basic and the Extended security options, which provide respectively successively more complete control over network access.

Tunneling

In its original meaning, the term *tunneling* was applied to the ability to allow foreign protocols to move through an IP network. More recently, the term has been applied to such tasks as:

- allowing remote modem connections to pass to a local host in a secure manner, as in Microsoft's implementation of the point-to-point tunneling protocol or PPTP

- passing multicast messages across non-multicast-enabled networks or network segments

IP Multicast

The IETF is currently actively working to specify standards that will define interoperability for the exchange of multimedia data across the Internet and within Intranets. When completed, these specifications will make up the *protocol independent multicast* or PIM standard. In addition, many networking hardware vendors, such as Cisco Systems, plan to implement PIM in such a way as to provide:

- bandwidth management
- interoperability with the MBONE
- security

Network Monitoring and Debugging

With today's complex, diverse network topologies, a router's ability to aid the monitoring and debugging process is critical. As the junction point for multiple segments, a router sees more of the complete network than most other devices. Many problems can be detected and even solved using information that routinely passes through the router.

Effective IP routing implementations should provide a means to monitor:

- any routing metrics
- IP-related interface parameters, including whether the interface and interface physical layer hardware are up, whether certain protocols (such as ICMP and Proxy ARP) are enabled, and the current security level
- IP-related protocol statistics, including the number of packets and number of errors received and sent by the following protocols: IP, TCP, User Datagram Protocol (UDP), EGP, IGRP, Enhanced IGRP, OSPF, IS-IS, ARP, and Probe
- logging of all BGP, EGP, ICMP, IGRP, Enhanced IGRP, OSPF, IS-IS, RIP, TCP, and UDP transactions
- reachability information between nodes
- the active accounting database, including the number of packets and bytes exchanged between particular sources and destinations
- the contents of the IP cache, including the destination IP address, the interface through which that destination is reached, the encapsulation method used, and the hardware address found at that destination
- the current state of the active routing protocol process, including its update interval, metric weights (if applicable), active networks for which the routing process is functioning, and routing information sources

- the current state of the routing table, including the routing protocol that derived the route
- the next IP address to send to
- the number of intermediate hops taken as a packet traverses the network
- the reliability of the source node
- the router interface to use
- whether the network is directly connected
- whether the network is subnetted

APPENDIX B

WEB SERVER QUICK REFERENCE

This appendix addresses the two most widely used Web server applications: Apache and Internet Information Server. It consists of two tables, each of which outlines configuration characteristics that can affect the bandwidth consumption of these servers. Table B-1 deals with Apache. Table B-2 covers IIS.

Apache

See Table B-1 for an extensive listing of Apache configuration directives.

TABLE B-1

Apache Configuration Directives

Directive	Syntax	Default Setting	Context of Use	Description
AccessConfig	AccessConfig file-name	AccessConfig conf/access.conf	Server configuration; setting up a virtual host	The server will read this file for more directives after reading the ResourceConfig file. Filename is relative to the ServerRoot. This feature can be disabled using: AccessConfig /dev/null. Historically, this file only contained <Directory> sections; in fact it can now contain any server directive allowed in the server config context.
AccessFileName	AccessFileName filename file-name...	AccessFileName .htaccess	Server configuration; setting up a virtual host	When returning a document to the client the server looks for the first existing access control file from this list of names in every directory of the path to the document, if access control files are enabled for that directory. For example: AccessFileName .acl before returning the document /usr/local/web/index.html, the server will read /.acl, /usr/.acl, /usr/local/.acl, and /usr/local/web/.acl for directives, unless they have been disabled with <Directory/>.

AddModule	AddModule module module...	None	Server configuration	The server can have modules compiled which are not actively in use. This directive can be used to enable the use of those modules. The server comes with a preloaded list of active modules; this list can be cleared with the ClearModuleList directive.
AllowOverride	AllowOverride override override...	AllowOverride All	Directory access	When the server finds an .htaccess file (as specified by AccessFileName) it needs to know which directives declared in that file can override earlier access information. Override can be set to: None, to prevent the server from reading the file; All, to cause the server to allow all directives; AuthConfig, to allow use of authorization directives; FileInfo, to allow use of directives controlling document types; Indexes, to allow use of directives controlling directory indexing; Limit, to allow use of directives controlling host access; or Options, to allow use of directives controlling specific directory features.

TABLE B-1

Apache Configuration Directives (*Continued*)

Directive	Syntax	Default Setting	Context of Use	Description
AuthName	AuthName auth-domain	None	Directory management	This directive sets the name of the authorization realm for a directory. This realm is given to the client so that the user knows which username and password to send. It must be accompanied by AuthType and require directives, and directives such as AuthUserFile and AuthGroupFile to work.
AuthType	AuthType type	None	Directory management	This directive selects the type of user authentication for a directory. Only Basic is currently implemented. It must be accompanied by AuthName and require directives, and directives such as AuthUserFile and AuthGroupFile to work.
BindAddress	BindAddress saddr	BindAddress *	Server configuration	A UNIX http server can either listen for connections to every IP address on the server machine, or just one IP address of the server machine. Saddr can be *, an IP address, or a fully qualified Internet domain name. If the value is *,

then the server will listen for connections on every IP address; otherwise it will only listen on the IP address specified. Only one BindAddress directive can be used. For more control over which address and ports Apache listens to, use the Listen directive instead of BindAddress.

ContentDigest

CoreDumpDiretory

Directive	Syntax	Context	Description
ClearModuleList	ClearModuleList	Server configuration	The server comes with a built-in list of active modules. This directive clears the list. It is assumed that the list will then be repopulated using the AddModule directive.
ContentDigest on|off	ContentDigest off	Server configuration	This directive enables the generation of Content-MD5 headers. MD5 is an algorithm for computing a "message digest" (sometimes called "finger-print") of arbitrary-length data, with a high degree of confidence that any alterations in the data will be reflected in alterations in the message digest. Note that this can cause performance problems on your server since the message digest is computed on every request (the values are not

TABLE B-1

Apache Configuration Directives (*Continued*)

Directive	Syntax	Default Setting	Context of Use	Description
			cached). Content-MD5 is only sent for documents served by the core, and not by any module. For example, SSI documents, output from CGI scripts, and byte range responses do not have this header.	
CoreDumpDirectory directory	The same location as ServerRoot	Server configuration	This controls the directory to which Apache attempts to switch before dumping core. The default is in the Server-Root directory; however, since this should not be writable by the user the server runs as, core dumps won't normally get written. If you want a core dump for debugging, you can use this directive to place it in a different location.	DefaultType
DefaultType mime-type	DefaultType text/html	Server configuration; setting up a virtual host; directory management	There will be times when the server is asked to provide a document whose type cannot be determined by its MIME types mappings. The server must inform the client of the content-type of the document, so in the event of an	<Directory>

Directive	Default	Context	Description
			unknown type it uses the DefaultType. For example: DefaultType image/gif would be appropriate for a directory that contained many gif images with filenames missing the .gif extension.
<Directory directory>...</Directory>	None	Server configuration; setting up a virtual host	<Directory> and </Directory> are used to enclose a group of directives, which will apply only to the named directory and subdirectories of that directory. Any directive that is allowed in a directory context may be used. Directory is either the full path to a directory, or a wild-card string. In a wild-card string, '?' matches any single character, and '*' matches any sequences of characters. As of Apache 1.3, you may also use [ ] character ranges like in the shell. Also as of Apache 1.3 none of the wildcards match a '/' character, which more closely mimics the behavior of UNIX shells.
<DirectoryMatch>			

TABLE B-1

Apache Configuration Directives (*Continued*)

Directive	Syntax	Default Setting	Context of Use	Description
<DirectoryMatch regex>...</DirectoryMatch>	None	Server configuration; setting up a virtual host	<DirectoryMatch> and </DirectoryMatch> are used to enclose a group of directives which will apply only to the named directory and subdirectories of that directory, the same as <Directory>. However, it takes as an argument a regular expression.	DocumentRoot
DocumentRoot directory-filename	DocumentRoot /usr/local/apache/htdocs	Server configuration; setting up a virtual host	This directive sets the directory from which httpd will serve files. Unless matched by a directive like Alias, the server appends the path from the requested URL to the document root to make the path to the document. Example: DocumentRoot /usr/web then an access to http://www.myhost.com/index.html refers to /usr/web/index.html.	ErrorDocument
ErrorDocument error-code document	None	Server configuration; setting up a virtual host; directory management	In the event of a problem or error, Apache can be configured to do one of four things: output a simple hardcoded error message; output a customized message; redirect to a	ErrorLog

				<Files>
			local URL to handle the problem; redirect to an external URL to handle the problem/error. The first option is the default, while options 2–4 are configured using the ErrorDocument directive, which is followed by the HTTP response code and a message or URL.	
ErrorLog filename	ErrorLog logs/error_log	Server configuration; setting up a virtual host	The error log directive sets the name of the file to which the server will log any errors it encounters. If the filename does not begin with a slash (/) then it is assumed to be relative to the ServerRoot.	
				<FilesMatch>
<Files filename>... </Files>	None	Server configuration; setting up a virtual host	The <Files> directive provides for access control by filename. It is comparable to the <Directory> directive and <Location> directives. It should be matched with a </Files> directive. Directives that apply to the filename given should be listed within. <Files> sections are processed in the order they appear in the configuration file, after the <Directory> sections and .htaccess files are read, but before <Location> sections.	

TABLE B-1

Apache Configuration Directives (*Continued*)

Directive	Syntax	Default Setting	Context of Use	Description
<FilesMatch regex>...</Files>	None	Server configuration; setting up a virtual host	The <FilesMatch> directive provides for access control by filename, just as the <Files> directive does. However, it accepts a regular expression.	Group
Group unix-group	Group #-1	Server configuration; setting up a virtual host	The Group directive sets the group under which the server will answer requests. In order to use this directive, the stand-alone server must be run initially as root. Unix-group is one of: A group name, refers to the given group by name; # followed by a group number, refers to a group by its number. It is recommended that you set up a new group specifically for running the server. Some admins use user nobody, but this is not always possible or desirable.	HostNameLookups
Host-NameLookups on | off | double	Host-NameLookups off	Server configuration; setting up a virtual host; directory management	This directive enables DNS lookups so that host names can be logged (and passed to CGIs/SSIs in REMOTE_HOST). The value double refers to doing double-reverse DNS. That is, after a reverse lookup	IdentityCheck

Directive	Default	Context	Description	See also
			is performed, a forward lookup is then performed on that result. At least one of the ip addresses in the forward lookup must match the original address. (In "tcpwrappers" terminology this is called PARANOID)	
IdentityCheck boolean	IdentityCheck off	Server configuration; setting up a virtual host; directory management	This directive enables RFC1413-compliant logging of the remote user name for each connection, where the client machine runs identd or something similar. This information is logged in the access log. Boolean is either on or off	<IfModule>
<IfModule [!] module-name>...</IfModule>	None	Applies to all aspects of Apache configuration	The <IfModule test>...</IfModule> section is used to mark directives that are conditional. The directives within an IfModule section are processed only if the test is true. If test is false, everything between the start and end markers is ignored.	Include
Include filename	None	Server configuration	This directive allows inclusion of other configuration files from within the server configuration files.	KeepAlive

TABLE B-1

Apache Configuration Directives (*Continued*)

Directive	Syntax	Default Setting	Context of Use	Description
KeepAlive max-requests (Apache 1.1) or KeepAlive on/off (Apache 1.2)	KeepAlive 5 (1.1) or KeepAlive On (1.2)	Server configuration	This directive enables KeepAlive support. In Apache 1.1, it sets max-requests to the maximum number of requests you want Apache to entertain per request. A limit is imposed to prevent a client from hogging your server resources. Set this to 0 to disable support. In Apache 1.2 and later set to "On" to enable persistent connections, "Off" to disable.	KeepAliveTimeout
KeepAliveTimeout seconds	KeepAliveTimeout 15	Server configuration	The number of seconds Apache will wait for a subsequent request before closing the connection. Once a request has been received, the timeout value specified by the Timeout directive applies.	<Limit>
<Limit method method...>... </Limit>	None	Can be applied to any area of Apache configuration	<Limit> and </Limit> are used to enclose a group of access control directives, which will then apply only to the specified access methods, where method is any valid HTTP method. Any directive except another <Limit> or <Directory> may be used; the majority will be unaffected by the <Limit>.	Listen

Directive	Default	Context	Description
Listen [IP address :] port number	None	Server configuration	The Listen directive instructs Apache to listen to more than one IP address or port; by default it responds to requests on all IP interfaces, but only on the port given by the Port directive. Listen can be used instead of BindAddress and Port. It tells the server to accept incoming requests on the specified port or address-and-port combination. If the first format is used, with a port number only, the server listens to the given port on all interfaces, instead of the port given by the Port directive. If an IP address is given as well as a port, the server will listen on the given port and interface.
ListenBacklog backlog	ListenBacklog 511	Server configuration	The maximum length of the queue of pending connections. Generally no tuning is needed or desired, however on some systems it is desirable to increase this when under a TCP SYN flood attack.

ListenBacklog

<Location>

TABLE B-1

Apache Configuration Directives (*Continued*)

Directive	Syntax	Default Setting	Context of Use	Description
\<Location URL\>...\</Location\>	None	Server configuration; setting up a virtual host	The \<Location\> directive provides for access control by URL. It is comparable to the \<Directory\> directive, and should be matched with a \</Location\> directive. Directives that apply to the URL given should be listed within. \<Location\> sections are processed in the order they appear in the configuration file, after the \<Directory\> sections and .htaccess files are read.	\<LocationMatch\>
\<LocationMatch regex\>...\</LocationMatch\>	None	Server configuration; setting up a virtual host	The \<LocationMatch\> directive provides for access control by URL, in an identical manner to \<Location\>. However, it takes a regular expression as an argument instead of a simple string.	LockFile
LockFile filename	LockFile logs/accept.lock	Server configuration	The LockFile directive sets the path to the lockfile used when Apache is compiled with either USE_FCNTL_SERIALIZED_ACCEPT or USE_FLOCK_SERIALIZED_ACCEPT. This directive should normally be left at its default value. The main reason for	MaxClients

Directive	Example	Context	Description	
MaxClients number	MaxClients 256	Server configuration	changing it is if the logs directory is NFS mounted, since the lockfile must be stored on a local disk. The PID of the main server process is automatically appended to the filename. The MaxClients directive sets the limit on the number of simultaneous requests that can be supported; not more than this number of child server processes will be created.	MaxKeepAliveRequests
MaxKeepAlive Requests number	MaxKeepAlive Requests 100	Server configuration	The MaxKeepAliveRequests directive limits the number of requests allowed per connection when KeepAlive is on. If it is set to "0," unlimited requests will be allowed. We recommend that this setting be kept to a high value for maximum server performance.	MaxRequestsPerChild
MaxRequestsPer-Child number	MaxRequestsPer-Child 0	Server configuration	The MaxRequestsPerChild directive sets the limit on the number of requests that an individual child server process will handle. After MaxRequestsPerChild requests, the child process will die. If MaxRequestsPerChild is 0, then the process will never expire.	MaxSpareServers

TABLE B-1

Apache Configuration Directives (*Continued*)

Directive	Syntax	Default Setting	Context of Use	Description
MaxSpareServers number	MaxSpareServers 10	Server configuration	The MaxSpareServers directive sets the desired maximum number of idle child server processes. An idle process is one which is not handling a request. If there are more than MaxSpareServers idle, then the parent process will kill off the excess processes.	MinSpareServers
MinSpareServers number	MinSpareServers 5	Server configuration	The MinSpareServers directive sets the desired minimum number of idle child server processes. An idle process is one that is not handling a request. If there are fewer than MinSpareServers idle, then the parent process creates new children at a maximum rate of 1 per second.	NameVirtualHost
NameVirtualHost addr [: port]	None	Server configuration	The NameVirtualHost directive is a required directive if you want to configure name-based virtual hosts. Although addr can be hostname it is recommended that you always use an IP address; e.g., NameVirtualHost 111.223.44 with the NameVirtualHost directive the address to which your name-based virtual host names resolve. If you have multiple name-based hosts on	Options

Directive	Default	Description
		multiple addresses, repeat the directive for each address. Optionally you can specify a port number on which the name-based virtual hosts should be used; e.g., NameVirtualHost 111.22.33.44:8080.
PidFile		Server configuration; setting up a virtual host; directory management
Options [+]option [+]option...	None	The Options directive controls which server features are available in a particular directory. Option can be set to: None, to provide no extra features; All, to permit all options except for MultiViews; ExecCGI, to enable execution of CGI scripts; FollowSymLinks, to allow the server to follow symbolic links in this directory; Includes, to allow server-side includes; IncludesNOEXEC, to permit server-side includes but disable the #exec command and #include of CGI scripts; Indexes, to cause the server to return a formatted listing of the directory; MultiViews, to permit content negotiated MultiViews; SymLinksIfOwnerMatch, to cause the server to follow symbolic links only if the target file or directory is owned by the same user id as the link; and more.

TABLE B-1

Apache Configuration Directives (*Continued*)

Directive	Syntax	Default Setting	Context of Use	Description
PidFile filename	PidFile logs/httpd.pid	None	Server configuration	The PidFile directive sets the file to which the server records the process id of the daemon. If the filename does not begin with a slash (/) then it is assumed to be relative to the ServerRoot. The PidFile is only used in stand-alone mode.
Port	Port number	Port 80	Server configuration	Number is a number from 0 to 65535; some port numbers (especially below 1024) are reserved for particular protocols. See /etc/services for a list of some defined ports; the standard port for the http protocol is 80. The Port directive has two behaviors, the first of which is necessary for NCSA backwards compatibility (and which is confusing in the context of Apache). In the absence of any Listen or BindAddress directives specifying a port number, a Port directive given in the "main server" (i.e., outside any <VirtualHost> section >) sets the network port on which the server listens. If there are any Listen or

BindAddress directives specifying :number then Port has no effect on what address the server listens at. In addition, the Port directive sets the SERVER_PORT environment variable (for CGI and SSI), and is used when the server must generate a URL that refers to itself (for example, when creating an external redirect to itself). In no event does a Port setting affect what ports a VirtualHost responds on, the VirtualHost directive itself is used for that.

| require | require entity-name entity entity | None | Directory management | This directive selects which authenticated users can access a directory. The allowed syntaxes are: require user userid userid..., to allow only the named users to access the directory; require group group-name group-name..., to allow only users in the named groups to access the directory; require valid-user, to permit all valid users to access the directory. |

TABLE B-1

Apache Configuration Directives (*Continued*)

Directive	Syntax	Default Setting	Context of Use	Description
ResourceConfig	ResourceConfig filename	ResourceConfig conf/srm.conf	Server configuration; setting up a virtual host	The server will read this file for more directives after reading the httpd.conf file. Filename is relative to the ServerRoot. This feature can be disabled using: ResourceConfig /dev/null. Historically, this file contained most directives except for server configuration directives and <Directory> sections; in fact it can now contain any server directive allowed in the server config context.
RLimitCPU	RLimitCPU # or 'max' [# or 'max']	Unset; uses operating system defaults	Server configuration; setting up a virtual host	This directive takes 1 or 2 parameters. The first parameter sets the soft resource limit for all processes and the second parameter sets the maximum resource limit. Either parameter can be a number, or max to indicate to the server that the limit should be set to the maximum allowed by the operating system configuration. Raising the maximum resource limit requires that the server is running as root, or in the initial startup phase.

Directive	Syntax	Default	Context	Description
RLimitMEM	RLimitMEM # or 'max' [# or 'max']	Unset; uses operating system defaults	Server configuration; setting up a virtual host	This directive takes 1 or 2 parameters. The first parameter sets the soft resource limit for all processes and the second parameter sets the maximum resource limit. Either parameter can be a number, or max to indicate to the server that the limit should be set to the maximum allowed by the operating system configuration. Raising the maximum resource limit requires that the server is running as root, or in the initial startup phase.
RLimitNPROC	RLimitNPROC # or 'max' [# or 'max']	Unset; uses operating system defaults	Server configuration; setting up a virtual host	This directive takes 1 or 2 parameters. The first parameter sets the soft resource limit for all processes and the second parameter sets the maximum resource limit. Either parameter can be a number, or max to indicate to the server that the limit should be set to the maximum allowed by the operating system configuration. Raising the maximum resource limit requires that the server is running as root, or in the initial startup phase.

TABLE B-1

Apache Configuration Directives (*Continued*)

Directive	Syntax	Default Setting	Context of Use	Description
Satisfy	Satisfy 'any' or 'all'	Satisfy all	Directory management	Access policy if both allow and require used. The parameter can be either 'all' or 'any'. This directive is useful only if access to a particular area is being restricted by both username/password and client host address. In this case the default behavior ('all') is to require that the client passes the address access restriction and enters a valid username and password. With the "any" option, the client will be granted access if they either pass the host restriction or enter a valid username and password. This can be used to password restrict an area, but to let clients from particular addresses in without prompting for a password.
ScoreBoardFile	ScoreBoardFile filename	ScoreBoardFile logs/apache_status	Server configuration	The ScoreBoardFile directive is required on some architectures to place a file that the server will use to communicate between its children and the parent. The easier way to find out if your architecture requires a scoreboard file is to

Directive	Syntax	Default	Context	Description
				run Apache and see if it creates the file named by the directive. If your architecture requires it, then you must ensure that this file is not used at the same time by more than one invocation of Apache.
SendBufferSize	SendBufferSize bytes	None	Server configuration	The server will set the TCP buffer size to the number of bytes specified. Very useful to increase past standard OS defaults on high speed high latency (i.e., 100 ms or so, such as transcontinental fast pipes)
ServerAdmin	ServerAdmin email-address	None	Server configuration; setting up a virtual host	The ServerAdmin sets the email address that the server includes in any error messages it returns to the client. It may be worth setting up a dedicated address for this; e.g., ServerAdmin www-admin@foo.bar.com as users do not always mention that they are talking about the server.
ServerAlias	ServerAlias host1 host2...	None	Setting up a virtual host	The ServerAlias directive sets the alternate names for a host, for use with name-based virtual hosts.

TABLE B-1

Apache Configuration Directives (*Continued*)

Directive	Syntax	Default Setting	Context of Use	Description
ServerName	ServerName fully qualified domain name	None	Server configuration; setting up a virtual host	The ServerName directive sets the hostname of the server; this is used only when creating redirection URLs. If it is not specified, then the server attempts to deduce it from its own IP address; however, this may not work reliably, or may not return the preferred hostname. For example: Server-Name www.wibble.com would be used if the canonical (main) name of the actual machine were monster.wibble.com.
ServerPath	ServerPath path-name	None	Setting up a virtual host	The ServerPath directive sets the legacy URL pathname for a host, for use with name-based virtual hosts.
ServerRoot	ServerRoot directo-ry-filename	ServerRoot /usr/local/apache	Server configuration	The ServerRoot directive sets the directory in which the server lives. Typically it will contain the subdirectories conf/ and logs/. Relative paths for other configuration files are taken as relative to this directory.

ServerSignature	ServerSignature Off | On | EMail	ServerSignature Off	Directory management	
				The ServerSignature directive allows the configuration of a trailing footer line under server-generated documents (error messages, mod_proxy ftp directory listings, mod_info output,...). The reason why you would want to enable such a footer line is that in a chain of proxies, the user often has no possibility to tell which of the chained servers actually produced a returned error message. The Off setting, which is the default, suppresses the error line (and is therefore compatible with the behavior of Apache 1.2 and below). The On setting simply adds a line with the server version number and ServerName of the serving virtual host, and the EMail setting additionally creates a "mailto:" reference to the ServerAdmin of the referenced document.

TABLE B-1

Apache Configuration Directives (*Continued*)

Directive	Syntax	Default Setting	Context of Use	Description
ServerType	ServerType type	ServerType stand-alone	Server configuration	The ServerType directive sets how the server is executed by the system. Type is one of inetd. The server will be run from the system process inetd; the command to start the server is added to /etc/inetd.conf stand-alone. The server will run as a daemon process; the command to start the server is added to the system startup scripts. (/etc/rc.local or /etc/rc3.d/...) Inetd is the lesser used of the two options. For each http connection received, a new copy of the server is started from scratch; after the connection is complete, this program exits. There is a high price to pay per connection, but for security reasons, some admins prefer this option. Inetd mode is no longer recommended and does not always work properly. Avoid it if at all possible.
StartServers	StartServers number	StartServers 5	Server configuration	The StartServers directive sets the number of child server processes created on startup. As the number of processes is dynamically controlled depending on the load,

Directive	Syntax	Example	Location	Description
ThreadsPerChild	ThreadsPerChild number	ThreadsPerChild 50	Server configuration	there is usually little reason to adjust this parameter. When running with Microsoft Windows, this directive sets the total number of child processes running. Since the Windows version of Apache is multithreaded, one process handles all the requests. The rest are held in reserve until the main process dies.

This directive tells the server how many threads it should use. This is the maximum number of connections the server can handle at once; be sure to set this number high enough for your site if you get a lot of hits. |
| TimeOut | TimeOut number | TimeOut 300 | Server configuration | The TimeOut directive currently defines the amount of time Apache will wait for three things: total amount of time it takes to receive a GET request; amount of time between receipt of TCP packets on a POST or PUT request; amount of time between ACKs on transmissions of TCP packets in responses. |

TABLE B-1

Apache Configuration Directives (*Continued*)

Directive	Syntax	Default Setting	Context of Use	Description
User	User unix-userid	User #1	Server configuration; setting up a virtual host	The User directive sets the userid at which the server will answer requests. In order to use this directive, the stand-alone server must be run initially as root. Unix-userid is one of: A username, refers to the given user by name; # followed by a user number, refers to a user by their number. The user should have no privileges which result in it being able to access files that are not intended to be visible to the outside world, and similarly, the user should not be able to execute code which is not meant for httpd requests. It is recommended that you set up a new user and group specifically for running the server. Some admins use user nobody, but this is not always possible or desirable.

Internet Information Server

Internet Information Server 4.0 offers nearly two dozen modules. If you take the defaults when installing IIS 4, you may end up not only with unnecessary modules being loaded to your server, but also with many of those being started automatically at bootup, and therefore draining server resources.

Table B-2 gives a thumbnail sketch of 20 IIS modules, outlining what they do. All these modules can be included or omitted when you install IIS.

TABLE B-2

Internet Information Server 4.0 Modules

This IIS Module	Pertains to
Active Directory Service Interface	User interface to differing directory services (e.g., NT and OS/2)
Active Server Pages	Samples for Web publishing
ActiveX	Database access
Authentication Server	Authentication of remote user connection attempts; uses the RADIUS protocol
FrontPage Server Extensions	Web publishing
Index Server	Indexing Web and other content files
JScript	Web scripting
Mail Server	Mail management with SMTP
Microsoft Management Console	A unified format and structure for management information
News Server	Newsgroup analog to Mail Server
ODBC	Database access
Posting Acceptor	Web publishing
Script Debugger	Web scripting
Site Analyst	Mapping a Web site
SQL Server	Distributing database applications under NT/IIS
Transaction Server	Creating and managing custom distributed applications
Usage Analyst	A performance monitoring tool
VBScript	Web scripting
Web Publishing Wizard	Wizard-driven Web publishing
Windows Scripting Host	Scripting for the Web and for server administration

APPENDIX C

BANDWIDTH-FRIENDLY ALTERNATIVE SOFTWARE

Having spent so much ink describing the many ways a variety of software can gorge itself on your bandwidth, we thought it only fair to offer a ray of hope, in the form of a review of two examples of bandwidth-friendly software: the operating system FreeBSD and the browser Opera.

Although we ran no formal tests or benchmarks, we have loaded and worked with each of these, and can confirm that each lives up to its claims for speed of operation and ease of configuration.

FreeBSD

FreeBSD, like Linux, a freeware UNIX operating system, was designed for the Internet. FreeBSD includes what many consider the reference implementation for TCP/IP software, the 4.4 BSD TCP/IP protocol stack, thereby making it ideal for network applications and the Internet. FreeBSD also supports standard TCP/IP protocols.

Like most UNIX systems, FreeBSD allows you to:

- carry out remote SNMP configuration and management
- distribute network information with NIS
- resolve Internet hostnames with DNS/BIND
- route packets between multiple interfaces, including PPP and SLIP lines
- share files with NFS
- support remote logins
- transfer files with FTP
- use IP Multicast

What's more, FreeBSD includes a number of add-ons that facilitate your setting up a respectable Web server; among these are:

- SAMBA, to permit sharing file and print resources with Windows 95 and Windows NT

- a PCNFS authentication daemon, which provides support for machines running that suite

With the addition of an optional commercial package, FreeBSD can also support AppleTalk and Novell clients.

FreeBSD handles TCP extensions like the RFC-1323 high-performance extension and RFC-1644 extension for transactions, as well as SLIP and dial-on-demand PPP. Because of its stability and speed, a number of commercial organizations have adopted it as the operating system platform for their Web sites. Two of these are:

- Walnut Creek CD-ROM, whose home page appears in Fig. C-1, runs an FTP server that supports over 2500 simultaneous connections. This server is a single FreeBSD machine, which moves more than 7 terabits (that's 7 trillion bytes) of files every month.

- Yahoo Inc. one of the premier indexers of the Internet, also relies on FreeBSD for its Web servers.

FreeBSD includes kernel support for IP firewalling, as well other services, such as IP proxy gateways. Any 386 or higher PC running FreeBSD can act as a network firewall. Encryption software, secure shells, Ker-

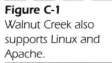

Figure C-1
Walnut Creek also supports Linux and Apache.

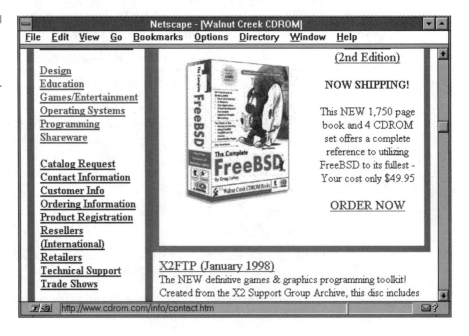

beros, end-to-end encryption, and secure RPC facilities are also available for FreeBSD.

Routing Under FreeBSD

As we learned earlier in this book, much of the bandwidth overhead a network can experience arises out of the need to route messages. For that reason, we briefly examine FreeBSD's routing mechanisms. We begin that process with Table C-1, which outlines example output of the command netstat -r.

In Table C-1 and the diagnostic output it represents, the first two lines specify the default and localhost routes respectively. The interface, shown in the column labeled Netif, for the local host is defined as lo0, a common designation for a loopback device. This entry therefore indicates that all traffic for localhost is internal, remaining within the host rather than being sent out over a LAN.

The next thing to note about this diagnostic are the *I0:e0:...* addresses, which are Ethernet hardware addresses. FreeBSD automatically identifies any hosts, such as test0, which are on a local network, and adds a route for each such host specifying that messages must travel over the indicated interface. In addition, FreeBSD associates a timeout, in the form of a parameter called *Expire,* with this type of route, which comes into play if a host is silent for the amount of time Expire represents. Should that happen, the route in question will automatically be deleted from the FreeBSD routing table.

TABLE C-1

Netstat Monitors
FreeBSD

Destination	Gateway	Flags	Refs	Use	Netif Expire
default	outside-gw	UGSc	37	418	ppp0
localhost	localhost	UH	0	181	lo0
test0	0:e0:b5:36:cf:4f	UHLW	5	63288	ed0
77	10.20.30.255 link#1	UHLW	1	2421	
host1	0:e0:a8:37:8:1e	UHLW	3	4601	lo0
host2	0:e0:a8:37:8:1e	UHLW	0	5	lo0

FreeBSD can also supply subnet routes. For example, in the preceding diagnostic output, 10.20.30.255 is the broadcast address for the subnet 10.20.30.

Under FreeBSD, both local network hosts and local subnets have their routes automatically configured by the routed daemon. If this daemon is not used, only static, explicitly defined routes will exist.

Finally, take a look at the column labeled Flags in Table C-1. Under FreeBSD, every route's characteristics are described by such flags, which include:

- U(Up): route is active
- H(Host): destination for the route is a single host
- G(Gateway): destination is a gateway to remote systems or networks
- S(Static): route was configured manually
- C(Clone): duplicates a route
- W(WasCloned): a route that was autoconfigured based upon a local area network (Clone) route
- L(Link): route involves references to Ethernet hardware

Default Routes

Under FreeBSD, if all known paths fail, a server has one last option: the default route. This is a special type of gateway route, usually the only one on the server, and always marked with a *c* in the Flags field. For hosts on a local area network, this gateway is set to whatever machine has a direct connection to the outside world.

A common configuration of a default route under FreeBSD might look like this:

```
[Local2] <-ether-> [Local1] <-PPP-> [ISP-Serv] <-ether-> [T1-GW]
```

In such a setup, the hosts Local1 and Local2 are at the local site, with the former being a PPP connection to an ISP. That ISP has, in the LAN at its site, the server to which you ultimately connect, as well as the hardware, a T1 line, through which you do so.

Given this configuration, Table C-2 summarizes the default routes for each of these machines.

	Host	Default Gateway	Interface
TABLE C-2 *Sample Default Routes Under FreeBSD*	Local2	Local1	Ethernet
	Local1	T1-GW	PPP

The T1-GW line, rather than the ISP machine to which it connects, was defined as the default gateway for the host Local1 because a PPP interface like that used in this example has an address on the ISP's LAN, which is also associated with this connection. Therefore, there is no need to send any additional traffic to the ISP server.

Dual Hosts

One other type of configuration that FreeBSD supports is that in which a host is associated with two different networks. In the strictest sense, any machine functioning as a gateway, such as Local1 in Table C-2, which uses a PPP connection to do so, constitutes what is referred to as a *dual-homed host*. But most frequently, this term refers only to a machine that actually participates in two LANs. This participation may take one of two forms. The machine may actually support two network adapters, each of which has an address on a separate subnet. Or, the machine may have only one adapter, but use aliasing based on the utility ifconfig.

In either case, FreeBSD routing tables are set up so that each subnet knows that its peer on the other is the defined gateway to that subnet. This configuration is most frequently used to implement packet filtering or firewall security in either or both directions.

Opera

The shareware browser *Opera*, whose point of origin you can see in Fig. C-2, combines several virtues, not least among them being its size and speed. Some features of this small, very fast browser are:

- Javascript support
- a hierarchical address list

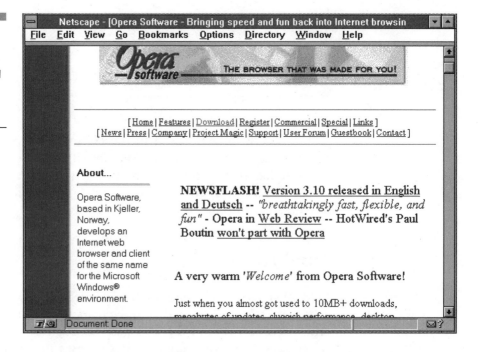

Figure C-2
As it loads, Opera's home page displays the message, central to any discussion of bandwidth, "less is more."

- support for Secure Socket Layer transmissions, including SSL 2.0 and 3.0, as well as TLS 1.0
- support for Netscape Plugins like:

  Adobe Acrobat
  Animated Widgets
  Calendar Quick
  Cosmo VRML player
  CPC View
  EchoSpeech
  Envoy
  QuickTime
  RealPlayer 5.0
  Shockwave Flash
  VDOLife

- keyboard shortcuts, such as Ctrl-R to reload a document
- cookies support
- news implemented with support for multiple news servers
- frames support

- multilevel docked hot list with import
- server Push (the more bandwidth-efficient push/pull mechanism)
- efficient cache handling (also, of course, significant in the context of bandwidth use)
- configuration settings for machines with little memory or hard disk (again relevant to any discussion of bandwidth, since, as we've seen, memory is the local resource that most closely interacts with communications)
- HTTP_REFERRER support
- support for Windows 95 and Windows NT

Installation and Requirements

In this section, we outline our experiences with downloading and configuring Opera.

Hardware Requirements.   As its Web site noted, less is more; Opera's platform requirements are so small as to almost seem unrealistic. We've outlined them in Table C-3, together with a comparison to our PC's characteristics, the latter given as a point of real-world reference.

Installing Opera.   Although the machine to which we downloaded and installed Opera is respectable, it's by no means cutting-edge. That, and the fact that the 1.1 MB, self-extracting and -installing Opera distribution executable downloaded in about 20 minutes across our 14.4 modem, and installed in about 37 seconds, is further evidence of the application's efficiency, and a further indication of its bandwidth-friend-

TABLE C-3	Component	Minimum	Recommended	Real-World
Requirements and Recommendations for Opera	Processor	386	486	Pentium 133
	RAM	6 MB	8 MB	32 MB
	Disk	2 MB free	4 MB free	About 500 MB free

Figure C-3
We accepted Opera's
default.

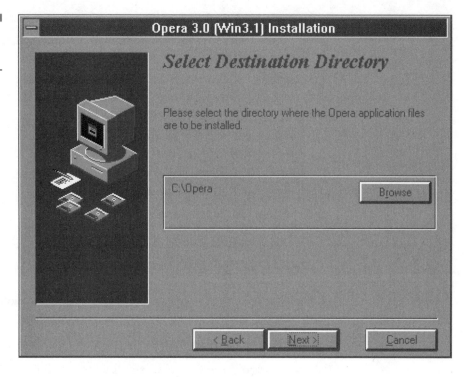

liness. Installing Opera was simplicity itself. From Windows (3.1—we're traditionalists), all we had to do was:

- select the downloaded executable with the Browse option of the Run command in Program Manager's File menu
- follow the on-screen instructions

Those instructions, as Fig. C-3 shows, began with a request for a destination directory.

Then, the installation routine, as Fig. C-4 illustrates, gave us the chance to make Opera our default browser. After a very few more such dialogs, Opera was installed. The entire process, from Welcome screen to "click Finish," took about a half-minute.

Configuring Opera. Opera required only one bit of configuration. When we first started the application, we received an error, but almost

Figure C-4
Being fond of
Netscape, we said no
to this inquiry.

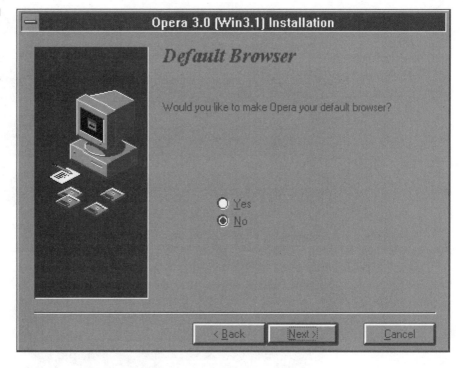

immediately, the solution to it. Opera told us that it could not run under the setting "Start with all files open," and provided a dialog through which we could correct the problem. Once we did so, the application ran beautifully. Nonetheless, we explored the browser's configuration options a little more, as a result. Those options, illustrated in Fig. C-5, allow you to tune the browser nicely.

The browser also has a context-sensitive, HTML-based help feature that clearly explains its configuration options, as you can see in Fig. C-6.

Winsock

In its Windows 3.x, that is, its 16-bit version, Opera requires a communications module that supports Windows Socket 1.1. If that isn't present, however, Opera can still function, but may need to have some of its options fine-tuned through the Advanced Preferences dialog shown in Fig. C-5.

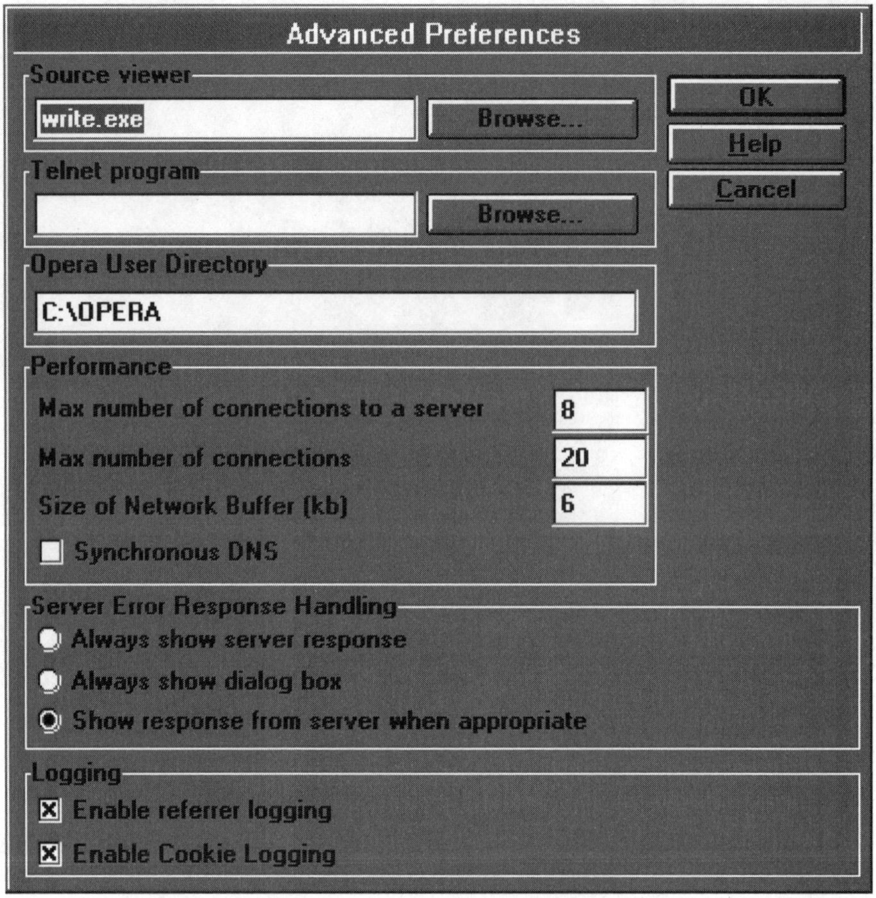

Figure C-5
Opera offers some
fairly sophisticated
configuration
options.

Table C-4 outlines some of the problems that can result from incorrectly or incompletely configured installations of Opera.

Electronic Mail

Opera supports sending, but not receiving, electronic mail. To send mail from within the browser, you must, as Fig. C-7 illustrates, configure the mail service, this time from the Mail option of the Preferences menu. Opera's documentation advises that addresses supplied to this dialog must conform to the SMTP standard. In addition, the

Figure C-6

Figure C-6
With this kind of
help, even a net-
working novice could
tailor Opera.

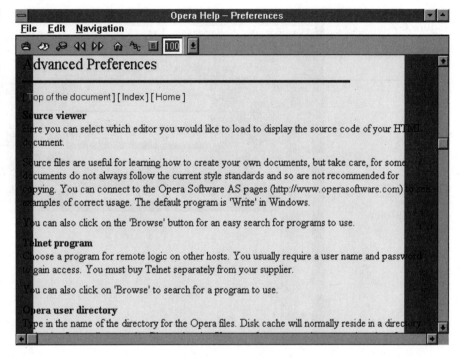

Opera Help – Preferences

File   Edit   Navigation

Advanced Preferences

[Top of the document] [Index] [Home]

Source viewer
Here you can select which editor you would like to load to display the source code of your HTML
document.

Source files are useful for learning how to create your own documents, but take care, for some
documents do not always follow the current style standards and so are not recommended for
copying. You can connect to the Opera Software AS pages (http://www.operasoftware.com) to see
examples of correct usage. The default program is 'Write' in Windows.

You can also click on the 'Browse' button for an easy search for programs to use.

Telnet program
Choose a program for remote logic on other hosts. You usually require a user name and password
to gain access. You must buy Telnet separately from your supplier.

You can also click on 'Browse' to search for a program to use.

Opera user directory
Type in the name of the directory for the Opera files. Disk cache will normally reside in a directory

application can, as Fig. C-7 also shows, act through an existing mail
package.

News.   As you would expect, Opera can also read news-groups. The
ability to do so, like the application's email arm, must be configured
through the Preferences menu.

TABLE C-4	**This Software**	**Can Cause**
Configuration-Related Problems in Opera	PC-NFS 5.0	errors if too big a network buffer is defined from within Opera. The suggested workaround for this problem is to set that buffer to 1 KB.
	Microsoft TCP/IP	Older versions of Microsoft TCP/IP require you to use the Synchronous Domain Name Service (DNS) when you run Opera. To set up this service, and an appropriate size for the network buffer, you must once again use the Advanced Preferences dialog to change the size of the network buffer.

Figure C-7
Opera provides two
ways to deal with
email.

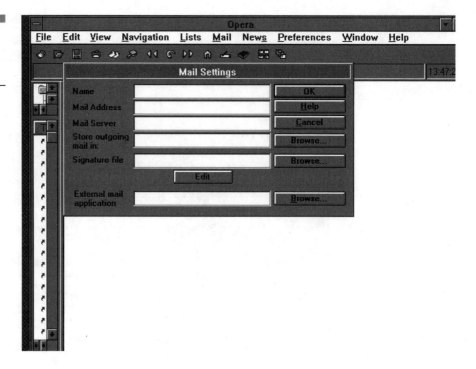

Using Opera

Running this browser, like the application itself, is quick and simple. We've outlined some of Opera's functional characteristics below.

Opera's main window, shown in Fig. C-8, consists of a menu bar, button bar, workspace with document windows, and a status line. The button bar provides fast access to the most used functions. Table C-5 outlines Opera's Button Bar.

Table C-6 summarizes Opera's status indicators. Figure C-9 shows those indicators in action.

Opera System Administration

Opera was developed with network use in mind. To simplify upgrading and to give the system administrator flexibility, a number of functions have been added. Probably the most important of these is the supersetup, contained in the opera.ini file, which controls all user selections

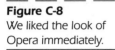

Figure C-8
We liked the look of
Opera immediately.

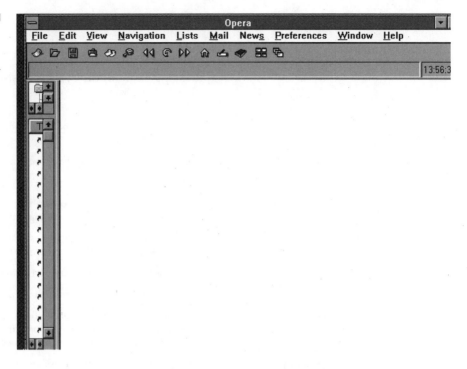

and preferences. To specify a file which contains operational parameters that can't be changed by users, that is, whose values will be valid only for a given session. and will thereafter be overridden, you must make the following entries in opera.ini.

```
[SYSTEM]
SUPER INI = <super-ini-filename>
```

Changing Preferences.   Since it can quickly become frustrating for the user if he or she creates a set of preferences that are constantly overriden, Opera allows the administrator to disable preference settings altogether. To do so, the administrator must make the following entries in either opera.ini or the defined super-setup file:

```
[MENU PREFS SETTINGS]
GENERIC = 1
BUTTON = 1
DOCUMENT = 1
ADVANCED = 0
FONT = 1
LINKS = 1
MAIL = 1
```

TABLE C-5	The Button	Handles
Opera's Button Bar	New	Opening a new, empty window.
	Open	Reading a file from disk. (Opera can display htm files and text files saved to disk.)
	Save	Saving the document in the active window to disk. By default, Opera will save the document in htm format so that you can later load it from disk.
	Print	Printing the document in the active window.
	Copy	Copying marked text.
	Search	Searching in the active window for text or a word. You can search both up and down; searching proceeds to the end of the document.
	Previous	Browsing backward in the document window's history.
	Reload	Rereading the document in the active document window.
	Next	Browsing forward in the document window's history.
	Home	Retrieving the document that has been selected as the home-page. If a homepage has been defined for the local window, it will be displayed; if not, the global homepage will be displayed.
	Direct	Keying in the address for a document to be retrieved. You can also type the address directly into the address field in a window. Whichever method you use, the document will be displayed in a new window, if "New window with direct addressing" has been selected in the General Preferences. Otherwise the document will be displayed in the active window, or another window if the active window is occupied.
	List	Opening the currently active subfolder of the hotlist. The top menu entries provide functionality for adding and retrieving bookmarks.
	Tile	Placing document windows so that they can all be displayed without overlap. It is possible to configure the program so that this is done automatically, that is, that space is made for new windows. This option is set in the Generic Preferences.
	Cascade	Positioning document windows to overlap.
	Graphics Control	By clicking on this button you can control whether graphics should be loaded and displayed, only loaded graphics should be displayed, or no graphics should be loaded and displayed. This option of course constitutes another bandwidth-friendly feature.

TABLE C-6	**This Indicator**	**Monitors**
Tracking Opera's Performance and Status	Document loading	How much of the document being fetched has been loaded. If the size is known, the progress is displayed in percentages. If the browser has no indication of document size, it displays the number of bytes or kilobytes that have been loaded. This indicator applies to documents of a variety of types, including those that contain graphics, sound, or other elements.
	Graphics loading	In the right of the two numbers it displays, how many extra elements such as linked or inlined graphics have been found in the document and remain to be loaded. With its left number, the indicator specifies how many such items have already been loaded.
	Generic loading	How much total data has been retrieved in the process of loading the document; includes the document itself, all figures, background, and even video clips, and so forth.
	Transfer speed	The current transfer speed of the document.
	Time elapsed since loading	The time elapsed since first requesting the document.
	Progress text	Whether Opera is connecting to a server, or loading data.
	History list	The addresses of downloaded documents.

```
NEWS = 1
VIEWER = 0
CACHE = 0
PROXY = 0
SOUND = 1
JPEG = 0
```

In this set of entries, 1 enables a user's ability to change preferences, whereas 0 disables that ability.

Location of Files. Normally all Opera setup files other than opera.ini are in the Opera directory. Underneath this directory is a cache directory. These default locations can be changed by setting the following values in opera.ini:

```
[USER PREFS]
OPERA DIRECTORY = <directory name>
HOT LIST FILE = <file name>
WINDOWS STORAGE FILE = <file name>
DIRECT HISTORY FILE = <file name>
CACHE DIRECTORY = <directory name>
```

Figure C-9
We liked Opera's
progress displays.

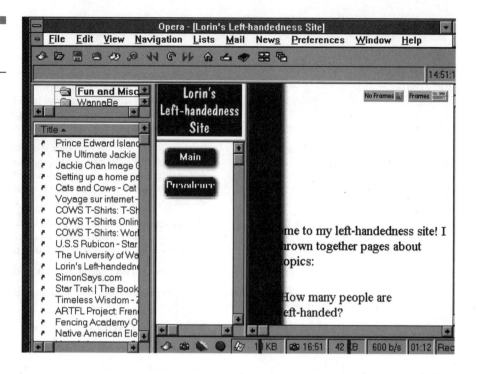

Opera ordinarily leaves its Window setup file, containing as it does information about display positioning, contents, history, and so on, accessible to users. However, a system administrator can override this, and ensure that the program always starts up with the same windows and contents.

The Direct addressing or history list is saved individually for each user. In similar fashion, the cache is specific to each user.

Opera has a system hot list, which makes it possible for the system administrator to have a generic hot list for all users, without users losing the possibility of keeping their own private lists. The system hot list is accessed from a menu option in the Navigate menu and under the List menu. What follows is an example of configuring the system hot list through opera.ini or super.ini.

```
[SYSTEM]
SYSTEM HOT LIST FILE = <filename>
```

Since individual users may wish to start Opera with different setups, the opera.ini/super.ini configuration scheme we've briefly investigated also makes it possible to use a setup file as a parameter at startup.

The Future of Opera

This final section provides a brief list of some of the anticipated enhancements to this user-, administrator-, and bandwidth-friendly browser.

Transparency to SOCKS. SOCKS is used to go securely through a firewall with authentication. Some Winsock implementations, like OnNet 4.0 from FTP software, allow an application to use SOCKS transparently. As yet, Opera does not do so, but as its designers put it, *that's on the list.*

Support for Java and CSS1. Opera support for both Java and CSS1 style sheets is planned for Version 4.0, due in the second quarter of 1998.

Receiving Email and Automatic Dialing. Opera's programmers are currently working on a separate email client, but have no specific anticipated completion date for this enhancement. Similarly, the application's designers have begun plans for an auto-dial feature.

APPENDIX D

BANDWIDTH TUNING
FROM THE GROUND UP

In this section, we attempt to provide you with a soup-to-nuts look at applying the principles we've presented. First, we offer a hypothetical network design that incorporates bandwidth efficiency. Then, we provide a set of tuning principles that can be applied to that design. Finally, we offer detail on tweaking operating system performance, by presenting a number of tricks from the HP-UX world.

Designing for Bandwidth Efficiency

In a study conducted by Cisco Systems, a design for a high-performance network, which would run in busy academic environments such as those of a college or university campus in an urban setting, was developed. That design assumed a number of characteristics for the current state of the networking environment:

- segments share a single hub and backbone
- routers and switches are present in heavy traffic areas
- user population consists of several thousand, mostly intermittent, users
- client hardware base consisting almost entirely of PCs
- communications medium almost entirely Ethernet, with a very small Token Ring presence
- protocol environment almost entirely TCP/IP, with some IPX/SPX, AppleTalk, NetBIOS presence
- a few centralized, network-level servers, as well as several workgroup-level servers
- traffic largely client/server, with an increasing amount of cross-subnet traffic

Among the goals of the Cisco design were to provide such environments with:

- the maximum bandwidth possible at every user port
- higher-speed client/server conversations
- the integration into the environment of such relatively new network players as Microsoft's Windows NT Server
- higher-speed workgroup servers, which can also act as multimedia servers
- ensuring that no single point of hardware failure will affect more than 100 users
- ensuring that the design may be fine-tuned based on monitoring of network traffic patterns
- ensuring that the design be as little constrained as possible by such circumstances as the physical location of wiring closets

The Physical and Network Access Layers

Given the nature of the postulated environment and its design goals, Cisco's study postulates the following as being needed at the hardware level in order to ensure high performance.

- 8000 user ports
- 100 users per wiring closet
- therefore, 80 desktop Ethernet switches

Addressing and Subnetting

In order to make the administration of IP addressing as simple as possible, that is, to minimize the cost of adding or otherwise modifying addresses, the Cisco study recommends the use of DHCP as the basis for address management.

As the Cisco researchers point out, DHCP is analogous to the plug-and-play features that gave the Apple Macintosh its initial, and ongoing, popularity. Assuming DHCP, therefore, the only addressing question that remains to be resolved is that of subnet size.

In the hypothetical campus environment as well as in so many others, average subnet size has gone from very large to rather small, as scalability of the network became a more prominent concern. Although the Cisco study acknowledges this fact of networking life, and therefore recommends no rule of thumb regarding subnet size, it does point out that this size must be easily adjustable, either up or down. The same concern—simplicity of administration—causes the study to recommend 255.255.255.0 as the environment's default subnet mask, thereby providing for as many as 254 hosts per subnet.

Switching

As we're all aware from experience and as the Cisco study points out, the number of building blocks, that is, of subnets, which a network design can propose is constrained by factors like building size and layout. Therefore, it is quite important to choose switching devices very carefully, and with an eye to providing both optimum traffic forwarding and redundancy, the latter so that, should a link between the segment and the network at large fail, another link can take over segment traffic.

Monitoring

Interestingly enough, the Cisco study emphasizes the need to design network monitoring techniques and tools into an environment from the ground up. The study emphasizes one area of network traffic—that between individual nodes' accessing the net and the workgroup or subnet servers that distribute material to such clients—as being particularly in need of tracking. As an example of the kind of fine-tuning that might be mandated by such observations, the study suggests the segregation of a few high-volume users onto a segment, and therefore a switch, of their own.

Workgroup Servers

It goes without saying that in a network, the demands made on servers, and particularly on workgroup servers, are most heavily influenced by the applications those servers distribute. It follows that monitoring network traffic resulting from such applications is key to physically and logically positioning such servers within a network in such a way as to optimize traffic flow.

Another similar decision, which can also be based at least in part upon an analysis of application-related traffic patterns, is the number of segments or subnets that should be configured into a workgroup server's NIC. Since, effectively, putting multiple subnets on a single NIC is functionally equivalent to having multiple NICs in a server, a commonsense, conservative rule of thumb would be to configure each workgroup server's NIC to manage an initial 6 to 10 subnets.

Routers

This level of connectivity hardware must provide:

- both inter- and intrasegment communications paths
- communications paths to enterprise servers and to a WAN and the Internet

Such paths can be provided in either of two ways, by:

- using stand-alone routers
- incorporating routing into existing switches, where possible

Of course, which of these you choose must depend upon constraints such as those imposed by physical plant, and by the nature of installed equipment.

In the Cisco study, both methods of routing deployment were addressed. In providing for stand-alone routers, the study predicates a total of four, each with up to eight Fast Ethernet adapters, and each offering two paths from the distribution layer, that is, from the workgroup-server level, to the core or enterprise-server level of the network. In the Cisco design, all routers within a network building block back each other up functionally, that is, provide redundancy in order to offer uninterrupted network flow.

Scaling Bandwidth Used by Traffic to and from the Core Layer

Again, Cisco postulates two means of increasing bandwidth available to conversations between the distribution and core layers in its hypothetical network. The first of these is simply to break existing subnets into smaller groups, and therefore introduce additional links. As it did in its discussion of routers, the study also points out that existing switches can sometimes be reconfigured to offer additional enterprise-to-workgroup bandwidth.

Enterprise Servers

Enterprise servers are those machines that are typically used by large numbers of users and traffic on which can encompass the entire network. Some commonplace examples include servers for:

- DHCP
- DNS
- email
- HTTP
- IP multicast

Because it is reasonable to assume that the near future of networking holds a steady increase in the demand for such enterprise-wide services and servers, the Cisco study cites the need for the use of such management tools as the Internet Group Management Protocol or IGMP.

Connecting to the Outside World

We've outlined below the technologies needed to provide not only connectivity of a network such as that on our hypothetical campus to the outside world, but also to legacy components such as the Token Ring and AppleTalk presences cited earlier in this chapter.

- access to legacy asserts of the network with routers at the core or enterprise level and translating bridging at the workgroup level
- redundant access to the Internet by means of multiple routers
- security for that access through an address translation firewall

Fine-Tuning Performance

Having examined some of the ways bandwidth efficiency can be designed into a network from the start, we turn, in this section, to discussing a few further ways to optimize network performance.

Limitations Imposed by Hardware

Any or all of the following hardware resources can, because of the physical limits in the quantity of data they can handle and the speed at which they can do so, can affect network throughput.

- CPUs and related, ancillary processors
- RAM
- virtual memory, that is, swap space on disk
- I/O devices
- network communications media and hardware

Earlier in this book, we looked at the effect of some of the limitations such components can place on bandwidth. For instance, we noted several times the critical role of the physical availability and proper configuration of server RAM. In the remainder of this section, we take a look at minimizing the negative effect on bandwidth of other of another critical server resource, its processors.

Fine-Tuning Processors

The first thing to note here is that, for PC server platforms at least, the presence of at least a math coprocessor, and ideally of multiple CPUs, can greatly improve server operating system performance and therefore, albeit indirectly, overall network performance.

Without a floating point coprocessor, a CPU must emulate, through software, the calculations such a processor would have performed. Such emulation is, as you might imagine, appreciably slower than doing the job with hardware.

Another processor-related influence on client/server processing and therefore traffic is the way in which direct memory access is carried out. Ideally, peripheral controllers should contain their own DMA bus chip, thereby relieving the DMA controller on the motherboard of at least some duties and traffic. In this context, you should note that some older peripheral controllers and SCSI host adapters cannot perform DMA.

Optimizing the Server OS

As Hewlett-Packard's own support staff points out on the H-P Web site, there is unfortunately no secret formula that will serve to fine-tune an operating system in every environment. H-P's techs acknowledge the commonsense dictum that the best settings for a given environment are defined by the types and volumes of data that environment must distribute.

Given this down-to-earth caveat, we review, in this section, some of the suggestions Hewlett-Packard makes for optimizing the performance of a server running HP-UX. Those suggestions address, not specific configuration settings, but rather system configuration parameters that most affect performance. We've selected those we feel most critical to network performance for inclusion here. So, this section discusses the effects on server and therefore network performance of:

- CPUs and other processors
- memory
- networking hardware
- swap space and other configurable OS kernel parameters

H-P points out four types of hardware components that directly and appreciably affect the performance of networked applications:

- CPU
- disk
- graphics
- RAM

Again relying on common sense, probably the most neglected and yet most valuable network management resource, H-P's techs note further that just as important as optimizing the performance of these components individually is the need to configure a system that will perform in a balanced way.

As so many of the sources we consulted in preparing this book have pointed out, and as we've noted throughout the book, truly effective configuration must be based upon:

- benchmarking that attempts to duplicate the intended types and patterns of use of a system and network

- monitoring of actual performance

With these precepts firmly in mind, let's turn to investigating some of Hewlett-Packard's suggestions for hardware configurations that will contribute to optimal server operating system performance.

CPUs

As we pointed out earlier in this chapter, many of the manipulations of data distributed by networked applications entail large numbers of floating point calculations, and CPU performance is the single most important factor in the efficiency with which such calculations are carried out.

Selecting a server (or client) CPU, the Hewlett-Packard Web site notes, is a tradeoff between cost, the nature and complexity of the calculations needed, and what you consider adequate performance. In other words, if a particular operation currently takes three seconds, is it worth it to spend several hundred or even thousand extra dollars on a processor upgrade that will allow you to perform the same task in 1.5 seconds?

Other characteristics that should be factored into the CPU fine-tuning equation are:

- the importance of calculations that the processor must perform
- the frequency with which these tasks must be carried out
- the impact of such calculations on overall application productivity

Having prioritized in this way the tasks a processor must handle, you're prepared to make the upgrade/don't upgrade decision, if you keep one last point in mind. Tasks most affected by CPU performance are those that involve more computation than disk access or graphics display.

Memory

Perhaps the most commonly asked configuration question, How much memory do I need? actually has two answers: As much as you need, and That depends.

Under operating systems like HP-UX and most other flavors of UNIX, which are virtual memory systems, determining the maximum amount of memory available to applications by the amount of swap space configured and not by the amount of physical memory. Insufficient physical RAM, therefore, will not necessarily prevent network applications from running under UNIX, but will have a serious negative impact on those applications' performance.

Memory and Swap Space

Given what we learned in the last section regarding how a virtual memory OS allocates memory to applications, it behooves us to explore further how such OSes work, in order to understand how to derive a blueprint for optimal memory configuration and management.

We begin with a quick summary of the nature of memory.

Memory 101. Memory is in fact and effect a form of high-speed data storage. Both the data and the instructions of any executing process are made available to a CPU by placing them in physical memory as the first stage of execution. That physical RAM is shared by all processes. Therefore, an operating system kernel relies on a per-process virtual address space that has been mapped into physical memory.

The term *memory management* refers to such mapping, as well as to parameters that govern:

- the configuration of physical and virtual memory
- the sharing of a system's memory resources by user-, application, and system-initiated processes

UNIX allows the aggregate total memory demands of user processes to exceed available physical memory by using the technique known as *demand paging.* Demand paging controls program execution by bringing into memory only those parts of a process that must be there at a given point in the program's execution, pushing back to disk those sections of the program that haven't run recently. HP-UX and other flavors of UNIX define a page as 4096 bytes.

Of course, however efficient paging might be, there remains no doubt that the more physical memory a server contains, the more data it can access and the more and larger processes it can execute without having to page. Keeping paging to a reasonable level is desirable; since paging involves disk I/O, it is slower than accessing RAM.

A last point to note about memory management is that not all physical memory is available to user processes. For instance, under HP-UX 10.20, the OS kernel always resides in main memory, never being swapped out. So, about 40 MB of server RAM is automatically occupied from the minute you boot such a machine.

Swap.   *Swap space*, that is, temporary, secondary storage that acts to augment physical memory, is initially allocated when a computer is configured. Such space can be made up of any of the following devices or portions of devices:

- an entire disk drive
- free space on a disk, outside the bounds of a file system
- a logical volume

UNIX also allows you to configure available free space within a file system as swap space. It's this form of swap that's known as a *swap file*. Using such files isn't the most efficient means of temporarily extending available memory, though; swapping by means of the file system is slower than doing so through a device or part of a device. If you must use file swapping, therefore, you should:

- consider it an interim solution only
- configure it with as high an execution priority as practical, in order to allow it to run as effectively as possible

Among the gazillion-plus cool things about UNIX is the fact that it allows you to add swap space on the fly to a running system. Both types of swap, device or file, may be implemented in this way. To add device swap to a machine that's up and running, you must:

- create a new file system on an unused disk
- define that file system as occupying only part of the disk
- define the unused space outside the file system as device swap

One last, and important note about swap space. Once it's activated, you must reboot a UNIX system to remove it.

Virtual Address Space.   Virtual memory, in order to map to processes, uses a technique called *virtual address space*. Such space houses information and pointers to the memory processes can reference. Under UNIX, there is one virtual address space for every process; these spaces have several functions:

- gives an overall description of the process
- holds pointers to what are known as per-process regions or *pregions*; these are another aspect of memory management under UNIX. Pregions are logical constructs that point to specific segments of a process, such as code, data, the kernel stack, the user stack, and shared-memory segments
- tracks pregions most recently involved in page faults

Under HP-UX, for instance, every process executes within a 4-gigabyte virtual address space. The structure that defines this space points to the process's own per-process region.

Any flavor of UNIX allows you to configure the kernel to control memory segment allocation. Table D-1 summarizes the types of segments whose assigned memory you might want to consider reworking, in order to improve server performance.

TABLE D-1

Configuring Memory to Improve Server Performance

This Segment	Holds	and Can Be Configured Through
text	A process's executable code; may be shared by multiple processes.	The kernel parameter *maxtsiz*
data	A process's data structures, whether initialized or uninitialized; can grow as needed by a program's logic.	The kernel parameter *maxdsiz*
stack	Information needed to dynamically allocate memory for such entities as local variables and subroutine return addresses. The kernel stack segment, contained within another called the u_area segment, holds a process's run-time stack when a machine is running in kernel mode. The size of both the u_area and the kernel stack segments is fixed, and cannot be reconfigured.	For stacks other than the u_area and kernel segments, the parameter *maxssiz*

Physical Memory and Performance

In a sense, there is no such thing under UNIX as too little physical memory. The amount of memory available to applications is defined by the amount of swap space configured, not by actual physical RAM present. But the latter does determine how much paging will be done while applications run. And as we've mentioned, paging carries with it a performance handicap. Because it moves data between physical memory and swap space, paging is slower than simply accessing physical memory. What's more, the more time a system spends paging, the slower it will execute applications.

There is a drop-dead level, so to speak, consisting of an amount of actual RAM. If a machine has less than this amount, it will spend almost all its CPU time paging, a condition known in data processing slang as *thrashing*. A system that has trouble executing even simple commands like cd or pwd in anything like an acceptable interval may be beginning to thrash. Clearly, such a condition requires immediate attention.

In the best of all worlds, of course, all operations would be done in physical memory, and paging would never be needed. But few of us work in such a world. So, in the remainder of this section, we discuss how to arrive at a balance between the amount of RAM you'd like to have, and what you might be able to afford.

As a means of helping us understand the question of minimum acceptable memory configurations, we introduce the Hewlett-Packard application Pro/ENGINEER. This package has many of the characteristics that you should consider; it helps to understand how memory is consumed.

On a workstation used as a dedicated Pro/ENGINEER platform, H-P documentation cites these types and amounts of memory consumption as typical.

- HP-UX: 40 MB
- windowing system: 6 to 8 MB
- Pro/ENGINEER: 29 MB

Of course, any other processes or services running on such a station will also consume memory. Even assuming the absence of such processes, the bare-bones configuration just outlined takes up about 75 MB of RAM, of which 40 MB is utterly irretrievable for any other use, since the operating system cannot be paged.

How, then, can we arrive at a reasonable figure for an amount of physical memory that will ensure adequate system performance? H-P documentation suggests two ways to do so.

■ Run a series of timed benchmark tests, made up of a set of processes that are common in your environment, on systems with increasing amounts of physical memory, in order to estimate the effect of additional memory on those applications' execution.

■ Use a system monitoring tool that allows you to track amounts and levels of paging. If too much paging is going on, overall or under certain circumstances, you probably need more actual RAM.

When Pro/ENGINEER was used as a benchmark application, and performance was observed in terms of memory size versus the amount of time that proved to be needed to perform typical operations in the application, results indicated that performance improved on a fairly steep curve as memory size increased, but only up to a point. Beyond that point, additional physical memory did not significantly improve performance.

From these results, we can conclude that an ideal amount of physical memory is that which falls on such a breakpoint. If the amount of RAM present is less than the breakpoint, a system will perform sluggishly. If physical memory exceeds the amount indicated by the breakpoint, there's more RAM on board than is actually needed.

As mentioned, device swap outperforms file system swap. One H-P engineer referred to device swap on two disks as the ideal swap space configuration, citing the fact that, when device swap is interleaved in this way—that is, when a system switches back and forth between two disks to allocate swap—better performance results. As proof of this theory, a program was written that accessed every page of a 300 MB set of data. That program was run on a system configured with 256 MB of RAM. Here are the results.

■ With one swap device, the program executed in 49:26 minutes.

■ With two swap devices, it ran in 31:06 minutes.

In other words, adding a second swap device resulted in almost a 40 percent reduction in run time.

To see how much swap space is configured on your system, and how much is in use, execute the command

```
/usr/bin/swapinfo -t
```

Figure D-1
This example of
swapinfo's output
was taken on a
Hewlett-Packard
9000/816, running
HP-UX 10.2.

```
# swapinfo
              Kb        Kb        Kb    PCT  START/        Kb
TYPE       AVAIL      USED      FREE   USED  LIMIT  RESERVE  PRI
dev       262144     17596    244548    7%      0       -    1
reserve        -     62672    -62672
memory     96788     50236     46552   52%
#
```

As you might imagine, you must be at root to execute this command, since swapinfo must open the kernel memory file /kmem to read swap usage information. Since this is a critical operating system file, access is usually restricted to root only. swapinfo will display information like that depicted in Fig. D-1.

Now we'll turn to Table D-2 to dissect this information.

If we run swapinfo with the -t option, we get results like those in Fig. D-2. As such output indicates, truly optimal swap space configurations can be arrived at only by monitoring swap usage with real data.

Disk

Several things can be done to improve disk I/O rates, and thereby, albeit indirectly, overall system and network performance. Table D-3 outlines some of these enhancements.

Checklists

We close this appendix with two tables. The first, Table D-4, offers in effect a checklist for configuration characteristics you need to keep in mind while attempting to fine-tune node and network performance in a busy, high-volume environment. The second, Table D-5, presents some tools, available under many flavors of UNIX, which allow you to monitor network server configuration and its effects on overall network performance.

TABLE D-2	**The Column Labeled**	**Indicates**
Interpreting Output from swapinfo	TYPE	That this system is configured with: ■ device swap ■ paging space reserved for processes that might need it ■ a paging area, sometimes called *pseudo-swap,* which represents the amount of system memory that can be used to hold pages, should all other areas of paging be in use
	Kb AVAIL	That this system has: ■ over 262 MB of device swap ■ no reserved paging space ■ nearly 97 MB of pseudo-swap
	Kb USED	That this system has: ■ over 17 MB of device swap in use ■ over 62 MB of reserved paging space in use ■ over 50 MB of pseudo-swap in use
	Kb FREE	That this system has: ■ over 244 MB of device swap available ■ no reserved paging space available ■ over 46 MB of pseudo-swap available
	PCT USED	That this system is using 7% of its device swap and 52% of its pseudo-swap.
	START/LIMIT	The block address, on the device used for paging, of the start of the paging area, and the number of blocks that can be used for paging, respectively.
	Kb RESERVE	The number of file system blocks that may be used for paging by processes initiated by ordinary users. On systems like this one, which are configured as solely device-swapping, this parameter is always null.
	PRI	The priority at which space in the swap area is used. Can be a value between 0 and 10; the lower the value, the higher the priority of swapping.

Figure D-2
The -t option adds file
system names to
swapinfo's output.

```
# swapinfo -t
               Kb        Kb        Kb    PCT  START/        Kb
TYPE         AVAIL      USED      FREE   USED LIMIT  RESERVE  PRI  NAME
dev         262144     17596    244548    7%     0        -    1  /dev/vg00/lvol:
reserve          -     59772    -59772
memory       96788     50224     46564   52%
total       358932    127592    231340   36%     -        0    -
#
```

TABLE D-3

Improving Disk I/O

This Step	Has These Effects
Upgrading to fast-wide SCSI-II	The fast-wide SCSI-II bus has a burst transfer rate of 20 MB/sec, compared to 5 MB/sec for the single-ended SCSI-II bus. The single disk data transfer rate for a 1.0 GB single-ended SCSI-II disk is 2.7 MB/sec, compared to 5.5 MB/sec for a 2.0 GB fast-wide SCSI-II. The 4.0 GB drive can attain a transfer rate of 8.9 MB/sec.
Distributing processing across multiple disks	Disk I/O performance can be improved by distributing data across multiple disks. In a single-disk configuration, a single drive must handle operating system access, application software access, and paging. If these tasks were distributed across several disks, performance improvements would almost certainly result.
Distributing swap space across two disks	Device swap space may be interleaved , or distributed across two disks. Interleaving improves paging performance.
Enabling asynchronous disk I/O	By default most versions of UNIX use synchronous disk I/O. In other words, file-system activity must complete to the disk before the program that initiated it can continue; applications don't regain control of the CPU until physical I/O finishes, regardless of whether that I/O involves user or operating system data. But when UNIX is configured to write to disks asynchronously, data blocks are still written using synchronous I/O, but status information such as file size is written to disk at some later time. As a result, the I/O operation's parent process regains CPU control almost immediately. Run-time performance increases by as much as 25% for I/O intensive applications when all disk writes occur asynchronously. But don't start planning your I/O reconfiguration just yet. Before enabling asynchronous disk writes, ponder the fact that if a system using such writes crashes, recovery of all data might not be possible. Under HP-UX, asynchronous I/O is enabled by setting the value of the kernel parameter fs_async to 1 and disabled by setting it to 0. The latter is, as mentioned, the default.

TABLE D-4

Performance
Tuning Checklist

In This Area	Be Sure to
Swap	Configure device rather than file system swap.
NFS	Ensure that your NFS configuration is up to date.
Network configuration	Keep in mind factors like the size of messages; the number of connection requests per second; the number of connected processes and of concurrently connected processes; the number and locations of users; the maximum number of connections allowed; the ratio of inbound to outbound traffic.
Size of messages	Keep an eye out for extremely large documents traveling the network, since it is through these that significant negative impact on throughput takes place.
Number of connection requests per second	Since, for example, 10,000 users checking mail once a day is easier to support than 1000 users checking mail every second during the day, keep close watch on the nature, number, and distribution of connection requests.
Number of connected processes	Since connected processes consume server memory, consider configuring the number of such connections your server will permit slightly lower than you might think necessary.
Number and location of users	1. Distribute users across drives. As the number of users per disk increases, the I/O to that disk increases nonlinearly, since searching a drive takes more time than searching other resources. 2. Distribute mailboxes across disks.

TABLE D-5

Tools for Monitoring UNIX Server Performance

This Tool	Can Be Used to
iostat	Provide information about disk I/O and CPU usage
lockstat	Give information on OS and application locking
mpstat	Present statistics about each processor on the system
netstat	Provide statistics about network functions
nfsstat	Offer statistics about NFS client/server statistics
pmap	Give you a breakdown on how much memory a process is using
protocol	Monitor processes and threads
snoop	Monitor network traffic
vmstat	Provide statistics about process, virtual memory, disk, trap, and CPU activity
bloatview	Present, through X windows, information about virtual memory
dkstat	Provide information about disk I/O and CPU use
netstat -C	Provide real-time, full-screen network status data

APPENDIX E

TROUBLESHOOTING TCP/IP

This appendix consists entirely of a single table that outlines a few of the problems most commonly experienced by TCP/IP, and possible solutions for those problems.

This Symptom	Can Be Caused by	and Can Be Corrected by
Inability to deliver datagrams.	Misconfigured or missing default routes.	1. View the host's routing table and look for entries labeled Default, which indicate default routes. 2. Ensure that such entries point to routers that control routes to remote hosts. 3. If there is no default route entry, establish one.
Inability to deliver datagrams.	DNS is not configured to do reverse lookups.	If the DNS server is not configured to perform reverse lookups, such attempts by end systems will time out. 1. Use the command appropriate to your operating system to determine the IP address of the node or host that has been involved in failed reverse-lookup attempts. 2. Try to connect to the indicated host or node by name and by IP address. 3. If the name-based connection attempt fails, but the address-based attempt succeeds, the system's DNS host table might be incomplete. 4. Ensure that appropriate address-to-hostname mappings exist in this table for every host on the network.
Connection attempts by some applications are successful, but attempts by others fail.	Missing or incorrectly configured router access lists.	1. Check each router in the path to determine if access lists have been configured for the router.

This Symptom	Can Be Caused by	and Can Be Corrected by
		2. If such lists exist, disable them.
		3. After disabling all access lists, check to see if formerly misbehaving applications function normally. If an application now runs properly, some access list is probably blocking traffic.
		4. To isolate the problem list, enable access lists one at a time until the application no longer functions.
		5. Having determined which access list is at fault, see if it filters traffic from any TCP or UDP ports. If the access list denies specific TCP or UDP ports, make sure that it does not deny the port used by the application in question.
		6. Reenable the modified list.
Hosts on one network can't communicate with hosts on another.	A default gateway being unspecified or incorrectly configured on either machine.	1. Determine whether local and remote hosts have a default gateway specification.
		2. If the default gateway specification is incorrect or missing, change or add a default gateway.
		3. Since you may have to reboot for this change to take effect, it may be preferable to specify a default gateway as part of the boot process.
Hosts on one network cannot access hosts on a different network. Error messages such as "host or destination unreachable" are encountered.	Routes are missing from the routing table.	Make sure IDs, addresses, and other configuration parameters are properly specified for the routing protocol you are using.
Performance for one or more network hosts is poor. Connections to these hosts are not established promptly.	Incorrectly configured DNS clients.	Check the DNS configuration file on clients. If the file is misconfigured, the client might wait until a query to one server times out before trying a second server, an NIS 1, or its host tables. This can cause excessive delays˜

GLOSSARY

In this section, we've tried to include every term directly related or significant to understanding bandwidth and how to make the most of it.

.au The filename extension of an audio file format, originally developed for Sun Microsystems, and later widely applied to audio files that are transmitted across the Internet.

14.4, 28.8, 56 Rates of transmission and colloquial names for modems; names refer to the number of kilobits per second the device can transmit.

802.x Shorthand referring to a group of standards for networks; best known: the 802.3 standard that defines Ethernet.

AC-3 An audio standard developed by Dolby Laboratories, which compresses six channels of digital audio into 384 Kbps.

ACK (Acknowledge) A character or signal indicating successful completion of a data communications or data processing task.

adaptive compression Data compression software that continually compresses the type and content of the data with which it works, and adjusts its algorithms to attempt to provide optimum compression.

address mask A pattern of bits sometimes used to separate an *Internet Protocol* or IP address into parts that represent, respectively, network, host computer, local resources, and subnet; also known as the *subnet mask.*

address resolution The translation of a symbolic address in domain-name format, such as *somebody@somewhere.com,* to the actual Internet Protocol (IP) address, such as 207.0.2.1; used to deliver transmissions; performed automatically by Domain Name Server applications.

administrative domain A section of the Internet or a local network under the control of one administrator or entity.

ADPCM (Adaptive Differential Pulse Code Modulation) An encoding method that stores audio signals in digital form.

ADSL (Asymmetrical Digital Subscriber Line) A technology for sending data at high speed over ordinary telephone lines.

agent   A program that performs a task for a user; especially, one that does so without or with minimal user intervention, for example: a Web crawler.

AIFF (Audio Interchange File Format)   File format sometimes used for sound files that must be transported across the Internet; requires a browser plug-in if it is to be "read" by the client.

alias   Substitute or nickname for a user or a process being executed by a computer.

analog   Relying on electromagnetic signals that, rather than occupying distinct states, represent a continuous range or an infinitely smooth curve of numeric values. Measurements that are characterized as analog include readings of voltage and current.

analog port   On some ISDN adapters, the equivalent of a modem.

anonymous FTP   To make a connection under the File Transfer Protocol (FTP) without having a user account on the remote host. Anonymous FTP accounts must be enabled and configured at the remote host before such sessions can be carried out.

Appletalk   Vendor-specific Apple Computer's data communications protocol.

ARPANET (Advanced Research Projects Agency Network)   A network, developed in the late 1960s for the Defense Department's Advanced Research Projects Agency, which became the basis for the Internet.

ASCII (American Standard Code for Information Interchange)   A method of encoding for computer data that uses seven or eight bits for data, and one for parity of error checking. As applied to data communications, a means of file transfer in which the data sent is formatted as ASCII.

ATM (Asynchronous Transfer Mode)   A new high-speed data communications technology whose links run at speeds from several hundred kilobits per second to several hundred megabits, and which features low transmission delays and dynamic allocation of bandwidth. Well suited to backbones, and to carrying real-time and multimedia data.

audio class   Software filters that deals with pulse-code-modulated data or analog signals.

AUI (Attachment Unit Interface)   That part of the Ethernet standard that specifies how a cable is to be connected to a Network Interface Card. AUI specifies a cable connected to a transceiver that plugs into a 15-pin socket on the network adapter.

authentication   The process, carried out automatically by appropriate software, of determining whether a remote user or computer has the right to access a local system.

automatic dial-up   The technique/configuration through which a connection to a remote host is made by a modem, automatically.

automatic forwarding   In email and other Internet applications, a configuration setting that causes messages to be forwarded to specified recipients automatically.

B-ISDN (Broadband Integrated Services Digital Network)   An ISDN service which employs broadband communications.

B-Router (Bridging Router)   A router that not only translates protocols and handles intersegment or internetwork addressing, but which also can connect segments or networks.

backbone   That part of a network's topology that constitutes the most functionally important segment. Other cable runs may be attached to a backbone.

bandwidth   In the strictest sense, the range of frequencies available on a given communications path for the transmission of data. In a more colloquial sense, synonymous with throughput, or the net data transfer rate of a network.

baseband   A means of data communication in which the electromagnetic signals that represent data are sent in their original form, rather than being modulated to a higher frequency or a different phase.

baud   In data communications, the number of times the medium's state or condition changes per second. For example, a 2400-baud modem changes the signal it sends on the phone line 2400 times per second. Each such change can correspond to more than one bit. Therefore, the actual bit rate of data transfer can exceed its baud rate.

binary transfer   A method of data transfer in which any bit pattern in the data moves through a connection. In other words, binary transfers are blind to data formatting, and simply move material as a continuous stream of bits, assuming that the receiver has the ability to decode that stream if such translation is needed.

BIND   UNIX implementation of the Domain Name System; short for *Berkeley Internet Name Domain.*

bind   To logically associate a hardware device with some characteristic, as in the Novell and Windows NT configuration tasks of binding a protocol or network adapter to a particular port.

BinHex   Short for *binary hexadecimal,* a nickname for a data file format. BinHex files combine data with information on resources such as program code, icons, and other information, and then translates the result into a string of 7-bit characters that can be treated as a text file.

bit   Short for *binary digit;* the numbering system and symbol set that contains only the digits 0 and 1, and which underlies all encoding of data for computers.

bit-mapped graphics   Images made up of matrices of pixels or dots. Bitmaps tend to lose clarity when transferred across networks.

block size   In the context of data communications, the size in bytes of the largest amount of data that can be transferred at one time. The larger the block size, the more efficient the transfer.

bounce   Jargon for the repeated transmission of a packet between the same points. Most frequently applied colloquially to email messages that are returned as undeliverable to their sender by an email server.

bpp (bits per pixel)   Number of bits used to represent the color value of each pixel in a digitized image.

bps (bits per second)   The most basic unit in which data transmission rates are measured. Similar to but not completely synonymous with *baud.* The convention is to use this term in lowercase.

BRA (Basic Rate Access)   In ISDN, the simplest form of service/access to a carrier.

broadband   A data communications link that can handle several channels of information, that is, ranges of frequencies, simultaneously. By extension, therefore, a very high-speed channel. Broadband functions by representing data as variations in high-frequency electromagnetic signals.

browser   A client application used to view and navigate between HTML documents, that is, Web pages. The most widely used browsers are Netscape Navigator and Microsoft Internet Explorer, but there are many others.

buffer An area of memory reserved to hold data which awaits processing or transmission.

cache A portion of RAM which is especially fast and reserved for the storage of data which must be accessed quickly or frequently.

CCITT (Comité Consultatif Internationale de Télégraphie et Téléphonie, or the Consultative Committee for International Telephone and Telegraph) An international standards organization dedicated to creating communications protocols that will enable global compatibility for the transmission of voice, data, and video across all computing and telecommunications equipment.

CERN The acronym for the French phrase *Congrèss Européen de Research Nucléaire*, which has come to be used as a nickname for the European Laboratory for Particle Physics in Switzerland. The site of origin of the http protocol and the HTML markup language, which together gave rise to the World Wide Web.

channel (1) A data communications pathway for a single stream of bits; (2) an area of a Web site to which a user can subscribe, and whose contents will thereafter be crawled and selectively downloaded automatically.

checksum A total, taken in binary arithmetic, which represents the characters in a packet, file, or area of memory. Used as a check for data correctness.

child (1) A computer process created automatically by another, previously running process; (2) an HTML document window created by or which results from a previously displayed window; and (3) a device attached to a system bus.

chmod Short for *change mode*, the UNIX command used to specify file access privileges for the owner of the file, members of the owner's user group, and all other users. Modes of access can include read, write, and execute permissions. Can be relevant to data communications processes that seek to access files on a UNIX system; without appropriate permissions, such attempts may fail.

CIX (Commercial Internet Exchange) An agreement among network providers that allows them to exchange commercial traffic.

Class A network Represented by values in the range 0 to 127 in the first part of an IP address; a network that can accommodate as many as 16 million hosts. A Class A network uses the first 8 bits of the 32-

bit IP address to represent the network number, and the remaining 24 bits for the host address.

client   A computer or application program that must rely on data and services from a larger system known as a server. In most cases, the client handles such tasks as user interface and local display, whereas the server carries out all significant or intensive processing, and then forwards the results of that processing to the client.

codec   Short for *coder-decoder;* hardware and/or software that converts such data file formats as graphics and audio to and from the forms needed for storage and transmission.

compatibility mode   An asynchronous, host-to-peripheral parallel port channel defined in the IEEE 1284-1944 standard. Compatible with existing peripherals that attach to the Centronics-style PC parallel port.

composite video   A signal that combines brightness, color, and synchronized video information into a single channel.

compressed video   A digital video image that has been processed to reduce the amount of data required to accurately represent its content.

compression   The translation of data, whether audio, digital, video, or some combination thereof, to a more compact form for storage or transmission.

connection   A negotiated method of communication between devices, whether carried out by hardware or software.

connectionless   A protocol that does not guarantee delivery of data.

connection-oriented   A protocol that attempts to guarantee delivery of data, and which has error detection and correction capabilities.

controllerless modem   A modem without the usual microcontroller, which relies on the host CPU for the AT command interpreter, control functions, and error checking.

CSMA/CD (Carrier Sense Multiple Access with Collision Detection)   The method used in TCP/IP networks for a station's receiving the ability to transmit. In effect, involves each node's listening at the transmission medium for the channel to become free, and then attempting to be the first to start talking. Under such a system, collisions can easily occur. When they do, transmitting stations wait for a predefined interval before attempting to gain control of the channel again.

CSU/DSU (Channel Service Unit/Digital Service Unit)   Any device that connects to a phone line at 56K/64K or T1, and provides a connection to network equipment.

cyclic redundancy checks (CRCs) A common data communications error-checking scheme.

daemon Pronounced like the word *demon*; a UNIX term for a program that runs automatically in the background without the need for any user intervention. Common daemons on UNIX systems include the print spooler daemon and the daemons that handle various protocols. For example, the httpd daemon is the UNIX server process that handles client requests for HTML documents.

datagram A data communications message or transmission that has been broken up into packets that are sent independently, and sometimes by different routes, and reassemble only when they reach their destination. Used in the TCP/IP protocol suite as well as many others. Enables transmissions to bypass bottlenecks.

DECnet Proprietary networking protocols used by Digital Equipment Corporation operating systems; not compatible with the Internet or TCP/IP.

dedicated line Also known as *leased line*; a permanently connected private telephone line between two locations. Often used to connect a local network to an Internet service provider.

DES (Data Encryption Standard) A model for encoding data so that reading it requires a software key. Sponsored by the United States government.

device Any circuit that performs a specific function; e.g., a serial port.

device ID A string that identifies a hardware device and distinguishes every logical device and bus from all others on a system.

dial-up A temporary network connection made over the phone system by means of a modem.

DIB (Device-Independent Bitmap) A file format designed to ensure that bitmap graphics created in one application can be loaded and displayed without any loss to or distortion of the image.

digital A method of representing data that relies on a signal made up of discrete numerical values.

digital signal processor An integrated circuit designed for high-speed data manipulations, used in audio, communications, image manipulation, and other applications.

digital video   A video signal represented as binary that describes a finite set of colors and brightness levels.

digitization   The process of transforming analog data or signals into digital form.

distributed   Term applied to applications or data that draw on the resources of more than one computer.

DLL (Dynamic Link Library)   Term coined by Microsoft for libraries of program modules that applications access through procedure calls.

DNS (Domain Name System)   The set of formats and procedures used to assign names to locations on the Internet and to correlate those names with IP network addresses.

domain   A named portion of the Internet or another network. The named entity must be under the control of a single person or organization. Can be represented by domain name suffixes like *.com,* indicating a commercial organization; *.org,* a nonprofit organization; and *.edu,* an educational institution.

domain name   Internet identifier that indicates domains, that is, networks under the jurisdiction of a particular individual or organization. Translated into an IP address when communications take place.

dot address   Nickname or alternate term for an Internet Protocol (IP) address. Derived from the fact that such addresses must be made up of four integers separated by periods or *dots.*

DPC (Deferred Procedure Call)   Method used in Windows NT for event scheduling.

DS0   Digital Signal Level 0 (64 Kbps).

DS1   Digital Signal Level 1 (T1).

DSVD (Digital Simultaneous Voice and Data)   The standard for network adapters, which are the ISDN equivalent of modems and which can transfer voice and data simultaneously.

dynamic IP address   Internet or other network address that is assigned "on the fly" from a pool of available addresses. Valid only for the duration of a user session.

ECP (Extended Capabilities Port)   An asynchronous, 8-bit-wide parallel channel defined by IEEE 1284-1944 that provides PC-to-peripheral and peripheral-to-PC data transfers.

Ethernet   In strictest terms, a LAN protocol designed by Xerox: baseband; uses a contention method like CSMA/CD; runs on coax or UTP; has a transmission rate range of 1 to 10 Mbps.

event set   A uniquely identified set of items about which a client can be notified.

Extended Industry Standard Architecture (EISA) bus   A 32-bit PC expansion bus designed as a superset of the Industry Standard Architecture (ISA) bus. EISA was designed to improve upon the speed and bandwidth of the PC expansion bus, while still supporting older ISA cards.

FDDI (Fiber Distributed Data Interface)   Standard for a type of data communications cabling that relies on strands of optical fibers rather than copper or other wires. Very nearly noise-free, and very high speed. Often used in backbones. The original FDDI specification called for a dual token ring structure that carried data at 100 megabits per second. Variations have been produced that offer a variety of data transmission rates.

filter   Software that selects data from a larger body according to specific parameters. In data communications, filters are often used to monitor transmission for particular protocols or points of origin, thereby forming the basis for firewalls.

firewall   Software and/or hardware, the latter usually in the form of routers that filter specific protocols or transmission sources, which attempts to provide a host computer with security by preventing specifically defined or unauthorized connections.

forms-based   Term applied to HTML documents that use the forms capability of HTTP and HTML to present information to or accept input from a user.

FQDN (Fully Qualified Domain Name)   An Internet or network domain name that includes all subsidiary domains up to the top level of the domain system, and so contains all the information needed to locate a site within the Internet or network. Analogous to a full path name in a computer file system.

frame   (1) In the context of Web pages, an area of a page that can be dealt with as if it were a separate HTML document, that is, independently filled, scrolled, and so on; (2) in the context of Internet video transmissions, a single complete video image. About 20 to 30 frames per second make up what appears to be full motion; and (3) in general data communications terms, synonymous with a single packet complete with control and address information. In serial data communications, a single character, including all data bits as well as any start, stop, or parity bits.

frame relay   A high-speed packet-transmission technology sometimes used for backbone connections, which assumes error checking and control will be handled by the systems at each end and therefore offers high throughput and low overhead.

FT1 (Fractional T1)   A fraction of a T1 that uses less than 24 DSOs.

FTP (File Transfer Protocol)   An application/member of the TCP/IP family of protocols used for transferring files between computers and, by extension, both the process of doing so and the system of programs that request and supply such transfers.

full duplex   Simultaneous bidirectional data flow.

gateway   A computer system that transfers data between normally incompatible applications or networks by reformatting data so that it is acceptable to its destination before passing it on.

Gbps (Gigabits per second)   A data transmission rate of one billion (1,000,000,000) bits per second. Available only in the most recent and highest-speed technology suites.

gopher   A menu-based system for searching Internet resources.

HCL (Hardware Compatibility List)   In Windows NT and 95, the extensive list of hardware products that have been tested and certified as compatible with these operating systems by Microsoft. Only items on this list can be guaranteed to run under these OSes.

IETF (Internet Engineering Task Force)   Volunteer group that investigates and solves Internet-related technical problems.

Industry Standard Architecture (ISA) Bus   An 8-bit, and later a 16-bit, expansion bus that provides a buffered interface from devices on expansion cards to the PC internal bus.

interrupt   An integer value used as a signal to a CPU that a hardware device wants to talk to its driver, and that the CPU should therefore temporarily put aside its current task in order to permit this conversation to take place.

interrupt-driven   Term applied to software, such as networking subsystems, which rely on interrupts to get and prioritize use of the CPU.

interrupt request   The method by which a hardware device requests service from its driver. A PC's motherboard uses a programmable interrupt controller or PIC to monitor the priority of all such requests. When a request is made, the CPU suspends whatever it's currently working on, and gives control to the device driver associated with the interrupt request number issued. The lower such a number is, for example, IRQ3, the higher the priority of the interrupt.

IPX (Internetwork Packet Exchange)   A transport layer protocol used in Novell networks. Analogous to IP in the TCP/IP suite.

IRP (I/O Request Packet)   In Windows environments, data structures in memory that drivers use to communicate with each other.

ISDN (Integrated Services Digital Network)   A network that relies on completely digital transmissions, media, and hardware, and which can simultaneously transfer a variety of types of information including voice, data, image, and video.

isochronous   Term applied to data communications that rely on slices of a CPU's processing time, rather than handshaking.

ISR (Interrupt Service Routine)   In Windows environments, a software routine that responds to a device's requesting services by issuing an interrupt.

JPEG   Image compression standard developed by the *Joint Photographic Experts' Group,* and used to pare down the size of image files to be transmitted across networks.

Kbps (Kilobits per second)   Data transmission rate of one thousand (1000) bits per second.

kernel   The core of the an operating system, which carries out its most basic tasks, such as process management, file management, hardware management, user authorization, and so on.

layered driver   In Windows environments, any of a collection of drivers that can be handled by the same ISR.

lossless compression   A category of file compression techniques that attempts to ensure that the original image will be presented at its destination without any loss of quality. Lossless compression therefore does not reduce image or other file sizes as radically as might otherwise occur.

lossy compression   A category of file compression techniques that does not attempt to ensure that images will retain their original quality upon being received and uncompressed. Many experts believe that up to 95 percent of the data in a typical image may be discarded without a noticeable loss in apparent resolution. Clearly, lossy compression can compact files more significantly than can lossless compression.

LRU (Least-Recently Used) An algorithm used to control swapping data or program instructions in and out of memory and CPU.

mail reflector An address from which mail received is automatically forwarded to a set of other addresses; software that forwards mail from such an address.

Mbps (Megabits per second) Data transmission rate of one million (1,000,000) bits per second; the rate currently accepted as adequate.

MPEG A video compression standard developed by the *Motion Pictures Experts' Group*.

mux. Nickname for a multiplexer, a networking hardware device that combines several data streams into a single one.

NDIS (Network Driver Interface Specification) An interface for network card drivers that provides transport independence for network cards.

NetBEUI (Network Basic Extended User Interface) Microsoft transport protocol used by stations running under older versions of Windows, that is, Windows 3.x. An enhancement of NetBIOS.

NetBIOS (Network Basic Input Output System) An older transport protocol. Developed by IBM.

NFS (Network File System) Set of protocols that allows you to use files on other network machines as if they were local. Originally developed by Sun Microsystems for UNIX, but currently in widespread use.

nibble mode An asynchronous, peripheral-to-host channel defined in the IEEE P1284 standard, which provides a channel through which the peripheral can identify itself to the host.

NIC (1) Network Information Center: Any organization responsible for supplying information about its network; (2) (Inter)NIC: the Network Information Center whose responsibility it is to coordinate such parameters as domain names for the entire Internet; (3) a Network Interface Card, an adapter through which a computer or other hardware device connects to a data transmission medium such as an Ethernet cable.

NMI (Nonmaskable Interrupt)   An interrupt that cannot be overruled by another request for service by the CPU.

NOC (Network Operations Center)   Those individuals and the location that are responsible for day-to-day network maintenance and management.

NREN (National Research and Education Network)   An effort by the United States to combine networks operated by different federal agencies into a single high-speed network.

octet   A term, usually used in the context of IP addresses, which refers to a single set of 8 bits, that is, to a byte.

OSI (Open Systems Interconnection)   A model for interoperability in data communications that defines seven layers of protocols, numbered in sequence from lowest, that is, closest to the network hardware, through highest or furthest removed from that hardware: Layer 1 (Physical), which defines the standards for transmission media; Layer 2 (Data Link), which packages data before it is transmitted; Layer 3 (Network), which transmits packets. TCP is a network layer protocol; Layer 4 (Transport), which routes packets. IP is a transport layer protocol; Layer 5 (Session), which provides services that permit the upper OSI layers to carry out such tasks as processing distributed applications and data; Layer 6 (Presentation), which defines the context or environment in which a user's network session will execute; Layer 7 (Application), which handles applications' interface to the network.

packet   The unit into which data communications messages are split before being forwarded separately and often by different routes to their destination. Packet sizes vary from as few as 40 to as many as 32,000 bytes, depending on network hardware and media. Most packets are less than 1500 bytes long.

parent   (1) A process that creates or *spawns* another process; (2) a hardware device such as a system bus that can have one or more children.

POP (Point of Presence)   A network or telephone company location that is closest to a customer or geographical area; analogous to a local post office.

port   (1) In the context of TCP/IP, a number that identifies a networked application. When data communications take place, packets contain information about the protocols they use and what application they're trying to communicate with. It's this application that is identified by a TCP/IP port number; (2) one of a computer's physical input/output channels.

POST   In terms of CGI or other processing of HTML documents, a method of accepting user input from a Web page.

PPP (Point-to-Point Protocol)   A protocol that establishes a link to a remote host, after which ordinary transport protocols like TCP, IPX, or NetBEUI access network resources.

PPTP (Point-to-Point Tunneling Protocol)   An expansion upon PPP that provides packet encryption.

protocol   A set of definitions for how computers will converse with one another. Protocol definitions range from how bits are placed on a wire to the format of an electronic mail message.

pull technology   In client/server applications, the client's specifically requesting, and controlling the presentation of, data from a server, without the need for user intervention, and at scheduled intervals.

push technology   In client/server applications, to send data to a client without the client's requesting it.

real time   An operating mode in which the results of data processing are returned promptly to the requestor; in the context of bandwidth, tasks like IP multicast.

registry   In Windows NT and 95 environments, a hierarchical database of all system resources, both hardware and software. Intended eventually to replace completely separate .INI, that is, initialization files.

resource conflict   The result of more than one device attempting to access and use the same, nonsharable resource at the same time. Such conflicts can cause hardware devices to be partially or wholly nonfunctional, or an entire computer to hang up.

RISC (Reduced Instruction Set Computing)   A microprocessor design that employs rapid processing of a relatively small set of machine-level instructions. RISC limits the number of instructions that are built into the microprocessor, but implements each so that it can be carried out very quickly—usually within a single CPU clock cycle, that is, one millionth (0.000001) of a second.

RLE (Run Length Encoding)   A data-compression technique in which successive bytes of identical data are converted to a 2-byte pair, consisting of the repeated data byte and the repeat count.

router   A hardware device that can transfer data between two networks that use the same protocols, but whose cabling schemes may differ.

sampling   The first step in the process of converting an analog signal into a digital representation. Measures the value of the analog signal at regular intervals, and then encodes these values into a digital representation.

SCSI (Small Computer System Interface)   An I/O bus designed as a method of connecting several classes of peripherals to a host system through a single physical port, and without the need for changes to software.

SCSI Configured AutoMatically   A protocol that defines a means of automatically setting the identifier of any SCSI devices connected to a computer's bus.

server   Software and/or hardware that makes resources available to another computer.

service provider   An organization that provides connections to a part of the Internet.

shell   That portion of a computer's operating system through which users interact with the kernel, and which acts as a command processor.

shell account   A type of interface on a dial-up connection in which you log in to a remote host and from that host use a command-line interface to gain access to the Internet.

SLIP (Serial Line Internet Protocol)   A protocol that uses IP over a standard telephone line and by means of a high-speed modem. SLIP is being superseded by PPP, but is still in common use.

SMTP (Simple Mail Transfer Protocol)   A widely accepted standard for the transfer of email.

SNMP (Simple Network Management Protocol)   The most widely used network management protocol.

SRI   A California-based research institute that runs the *Network Information Systems Center* (NISC), and which has played an important role in coordinating the Internet.

switched access   A network connection that can be created and destroyed as needed. Dial-up connections are the simplest form of switched connections. SLIP and PPP also often run over switched connections.

symmetrical compression   A compression system that uses equal amounts of processing resources for both compression and decompression. Used in applications in which both compression and decompression occur frequently, such as videoconferencing.

synchronous   Any operation that proceeds under control of a clock or timing mechanism.

T1   544 Mbps service (24 DSOs). Sometimes called *High-Cap, T-Span,* or *T-Carrier.*

T3   44.736 Mbps service (28 T1s).

TAPI (Telephony Application Program Interface)   A set of Win32-based application program interfaces that can be used to control modems and telephones. TAPI routines direct procedure calls from an application to the service-provider DLL for a modem.

TCP/IP (Transmission Control Protocol/Internet Protocol)   The protocol suite on which the Internet is based. TCP/IP is a connection-oriented reliable protocol.

TELNET   Short for *telephone network.* A terminal emulation protocol that allows you to log in to remote hosts across a network or the Internet, or an application that relies on the TELNET protocol.

time out   Event that can occur when two internal computer processes, or two data communications processes, must communicate. If, after a specified interval, one of the partners to the conversation fails to respond, the other will consider the conversation over.

TN3270   A special version of the TELNET program that communicates with IBM mainframes.

Token Ring   A network topology in which no member of the network has transmission privileges superior to those of the other members, and in which the ability to transmit is controlled through the passing of a software signal or token.

UART (Universal Asynchronous Receiver/Transmitter)   A component that contains both the receiving and transmitting circuits required for asynchronous serial communication.

UDP (User Datagram Protocol)   Another of the protocols upon which the Internet is based. UDP is a connectionless unreliable protocol.

UTP (Unshielded Twisted Pair)   Telephone-like cabling that relies on two strands of copper wire coiled about one another, but without significant exterior shielding.

UUCP (UNIX-to-UNIX copy) (1) A utility for copying files between UNIX systems, on which mail and USENET news services were built; (2) a form of Internet addressing derived from the uucp family of data communications utilities.

WAIS (Wide-Area Information Servers) A system for locating information in databases across the Internet. Most frequently used by libraries, particularly those of academic institutions.

X.400 An international massaging standard.

X.500 A standard for network directory services.

BIBLIOGRAPHY

In this section, we've included print and online sources of further information about bandwidth management. Among these are:

- sources that we drew upon to any significant degree in preparing this book
- sources that some of our sources in turn worked with

Print

Baird-Smith, Anselm, *Jigsaw: An object oriented server*, World Wide Web Consortium, February 1997. Source and other information are available at *http://www.w3.org/Jigsaw*.

Baker, Steve, "The Challenges of Integration," *WRQ, Inc.*, August 1996.

Bellovin, S. M. *Pseudo-Network Drivers and Virtual Networks; Procedure*, Winter USENIX Conference, Washington DC, pp. 229–244, 1990.

Berners-Lee, Tim, R. Fielding, H. Frystyk, "Informational RFC-1945—Hypertext Transfer Protocol—HTTP/1.0," MIT/LCS, UC Irvine, May 1996.

Boutell, T., T. Lane, et al., "PNG (Portable Network Graphics) Specification," W3C Recommendation, October 1996.

Braden, R., "Extending TCP for Transactions—Concepts," RFC-1379, USC/ISI, November 1992.

Braden, R., "T/TCP—TCP Extensions for Transactions: Functional Specification," RFC-1644, USC/ISI, July 1994.

Deignan, Michael, "Sharing Files Between NT and UNIX Systems," *Windows NT Magazine*, March 1997; *http://www.winntmag.com*.

Deutsch, L. Peter, Jean-Loup Gailly, "ZLIB Compressed Data Format Specification version 3.3," RFC-1950, Aladdin Enterprises, Info-ZIP, May 1996.

Deutsch, P., "DEFLATE Compressed Data Format Specification version 1.3," RFC-1951, Aladdin Enterprises, May 1996.

Edwards, Mark Joseph, "Samba: UNIX and NT Interoperability Made Easy," *Windows NT Magazine*, March 1997; *http://winntmag.com*.

Estrada, Susan, *Connecting to the Internet*, O'Reilly and Associates, 1994.

Farber, D. J., G. S. Delp, and T. M. Conte, "A Thinwire Protocol for Connecting Personal Computers to the INTERNET," RFC-914, University of Delaware, September 1984.

Fielding, R., J. Gettys, J. C. Mogul, H. Frystyk, T. Berners-Lee, "Hypertext Transfer Protocol—HTTP/1.1," RFC-2068, UC Irvine, Digital Equipment Corporation, MIT.

Heidemann, J., "Performance Interactions Between P-HTTP and TCP Implementation," *ACM Computer Communication Review,* 27(2) pp. 65–73, April 1997.

Hubley, Mary, "Advanced Server for UNIX: Integrating UNIX and Windows NT Servers," *Open Computing Magazine,* 1995; *http://www.att.com/unix_asu/whatis.html.*

Hummingbird Communications Ltd., "The Network File System: An Interoperable Distributed File System," 1996; *http://www.hummingbird.com/whites/nfswhite.html.*

Intergraph Software Solutions, "DiskAccess V3.0 Reviewer's Guide," January 1997.

Jacobson, Van, "Compressing TCP/IP Headers for Low-Speed Serial Links," RFC-1144, Lawrence Berkeley Laboratory, February 1990.

Jacobson, V., R. Braden, and D. Borman, "TCP Extensions for High Performance," RFC-1323, LBL and ISI; October 1988.

Jacobson, Van, "Congestion Avoidance and Control," *Proceedings of ACM SIGCOMM '88,* pp. 314–329, Stanford, CA, August 1988.

Jaegermann, Michael, "Samba Server HOWTO," January 1995; *http://zeus.chem.wvu.edu/misc/smb_serv/html/smb_se.html.*

Korb, J. T., "A Standard for the Transmission of IP Datagrams Over Public Data Networks," RFC-877, Purdue, September 1983.

Lie, H., B. Bos, "Cascading Style Sheets, level 1," W3C Recommendation, World Wide Web Consortium, 17 December 1996.

Manasse, Mark S., and Greg Nelson, "Trestle Reference Manual," Digital Systems Research Center Research Report # 68, December 1991.

McGregor, Glenn, "The PPP Internet Protocol Control Protocol," RFC-1332, May 1992.

Michael Santifaller, *Internetworking in a UNIX Environment,* second edition, TCP/IP and ONC/NFS: Addison-Wesley (Germany), 1994.

Mogul, J., "The Case for Persistent-Connection HTTP," Western Research

Laboratory Research Report 95/4, Digital Equipment Corporation, May 1995.

Mogul, Jeffery, Fred Douglis, Anja Feldmann, and Balachander Krishna-murthy, "Potential benefits of delta-encoding and data compression for HTTP," *Proceedings of ACM SIGCOMM '97*, Cannes, France, September 1997.

Nagle, J., "Congestion Control in IP/TCP Internetworks," RFC-896, Ford Aerospace and Communications Corporation, January 1984.

Nielsen, Henrik Frystyk, "Libwww—the W3C Sample Code Library," World Wide Web Consortium, April 1997.

Padmanabhan, V. N., and J. Mogul, "Improving HTTP Latency,"Computer Networks and ISDN Systems, v. 28, pp. 25–35, Dec. 1995. Slightly revised version in *Proceedings of the 2nd International WWW Conference '94:* Mosaic and the Web, October 1994.

Paxson, V., "Growth Trends in Wide-Area TCP Connections," *IEEE Network,* 8(4) pp. 8–17, July 1994.

Perkins, Drew D., "The Point-to-Point Protocol for the Transmission of Multi-Protocol Datagrams Over Point-to-Point Links," RFC-1171, Carnegie Mellon University, July 1990.

Perkins, Drew, and Russ Hobby, "The Point-to-Point Protocol (PPP) Initial Configuration Options," RFC-1172, Carnegie Mellon University and University of California/Davis; July 1990.

Postel, Jon B., "Transmission Control Protocol," RFC-793, Network Information Center, SRI International, September 1981.

Romkey, J. "A Nonstandard For Transmission Of IP Datagrams Over Serial Lines," RFC-1055, SLIP, June 1988.

Scheifler, R. W., J. Gettys, "The X Window System," *ACM Transactions on Graphics # 63,* Special Issue on User Interface Software.

Sharpe, Richard, "Just What is SMB?" 15 September 1996; *http://samba.anu.edu/cifs/docs/what-is-smb.html.*

Shearer, Dan, "History of SMB." 1996; *http://samba.anu.edu.au/cifs/docs/smb-history.html.*

Simpson, Bill, "The Point-to-Point Protocol (PPP) for the Transmission of Multi-Protocol Datagrams Over Point-to-Point Links," RFC-1331 May 1992.

Spero, S., "Analysis of HTTP Performance Problems," July 1994.

Sun Microsystems, "The NFS Distributed File Service," March 1995, *http://www.sun.com/sunsoft/solaris/desktop/nfs.html.*

Syntax Corporation, "TotalNET Advanced Server v5.0 White Paper." 1997; *http://www.syntax.com/totalnet/50wp/50whtpr.htm.*

Tanner, Jeff, "Common Internet File System," *WRQ, Inc.,* February 1997.

Touch, J., "TCP Control Block Interdependence," RFC-2140, USC/ISI, April 1997.

Touch, J., J. Heidemann, and K. Obraczka, "Analysis of HTTP Performance," USC/Information Sciences Institute, June 1996.

Waitzman, David, "A Standard for the Transmission of IP Datagrams on Avian Carriers," RFC-1149, April 1990.

Werden, Scott, "Integrating Windows NT into a UNIX Network," *WRQ, Inc.,* June 1997.

Wood, David, "SMB HOWTO," August 1996; *http://calvin.caltech.edu/docs/mdw/HOWTO/SMB-HOWTO.html.*

Cabling Standards

The ANSI/EIA/TIA-568-1991 Standard Commercial Building Telecommunications Wiring Standard.

UTP Connectivity Hardware: EIA/TIA Tech Sys Bulletin TSB-40A, December 1993 (Performance of Connectors and Patch Panels Above 20 MHz).

150-ohm Shielded Twisted Pair Cable and Data Connector Standards: EIA/TIA Tech Sys Bulletin TSB-53, 1992 (Type 1A cable).

Unshielded Twisted-Pair: EIA/TIA Tech Sys Bulletin TSB-36, November 1991 (Transmission Characteristics of Category 3-5 UTP cables).

Online

Information on	From	Can Be Found at
Balancing bandwidth consumption and video	Synthetic Aperture, a company involved in digital video	*www.synthetic-ap.com/ qt/internetvideo.html*
Balancing bandwidth consumption and well-designed Web pages	The Bandwidth Conservation Society	*www.infohighway.com/faster/index.html*

Information on	From	Can Be Found at
Basic Ethernet concepts	South Hills DataComm, a Pittsburgh-based VAR	*www.mazza.shillsdata.com/tech/Ethernet*
DEC's AllConnect for UNIX technologies	Digital Equipment Corporation	*www.unix.digital.com/tin/textit/ allconnect/acpaper2.htm*
Definitions of bandwidth- and processing-related terms	Microsoft	*www.unix.digital.com/tin/textit/ allconnect/acpaper8.htm*
Dial-up networking under Windows NT	Microsoft	*premium.microsoft.com/msdn/library/ books/platform/hdgwin95/d3/ sla50.htm*
Distributed memory (as featured in Chap. 10)	Institute for Computer Science of the Foundation for Research and Technology in Hellas, Crete	*support.microsoft.com/support/ kb/articles/q166/2/88.asp*
Documentation on the Cisco CPA1120 routing protocols	Cisco Systems	*www.csi.forth.gr/~markatos/ html_papers/TR190/paper.html*
Fast Ethernet	Cisco Systems	*alef0.cz/cprodocs/data/ciscopro/wan_sys/ 1100/protocol/igrp/ 1120ig01.htm*
Fast Ethernet	Students of the Worcester Polytechnic Institute	*www.cisco.com/warp/public/729/fec/ fefec_wp.htm*
FreeBSD (as featured in App. C)	FreeBSD Inc.	*www.freebsd.org*
General TCP/IP concepts	Cisco Systems	*www.cissco.com/warp/public/534/5.html*
Human factors in Web design	useit.com, a site dedicated to "usable information technology"	*www.useit.com/alertbox/9703a.html*
Improving performance of the HTTP protocol, in its 1.1 version	World Wide Web Consortium	*www.w3.org/Protocols/HTTP/ Performance/Pipeline.html*
ISDN	BellAtlantic	*www.bell-atl.com/isdn/consumer/ newshq.htm*
ISDN	Microsoft	*www.microsoft.com/windows/getisdn/ whatis.htm*
Lazy Receiver Processing (as featured in Chap. 7)	Rice University	*www.cs.rice.edu/CS/System/LRP*
Managing network queuing	Cisco Systems	*www.cisco.com/warp/public/731/ Protocol/dlsw5_rg.htm*
Managing network traffic flow	Cisco Systems	*www.cisco.com/warp/public/614/18.html*
Managing queues for network interfaces	Cisco Systems	*www.cisco.com/warp/public/614/16.html*

Information on	From	Can Be Found at
Material, originally published in *PC Week*, on managing and optimizing bandwidth consumption	Ziff-Davis online	*www5.zdnet.com/zdnn/content/ pcwk/1422/*
Net.Medic, a network monitoring tool	VitalSigns Software	*www.vitalsigns.com/products/nmp/ brochure/index.html*
NetScout, a network monitoring tool	NetScout Systems	*www.frontier.com*
Network monitoring tools featured in Chap. 15	Internet Engineering Task Force	*www.ietf.org*
Network monitoring tools for SunOs 4.1.x	Clemson University	*www.tiwet.clemson.edu/tiwet/sections/ gis/monitor.html*
Networking and other information on Windows 95	WinFiles.com, a Windows 95 shareware site	*www.windows95.com*
Networking concepts; network topologies; networks for academic environments	Asante Technologies, Inc.	*www.asante.com/education_primer/ intro.html*
Opera, a shareware browser (as featured in App. C)	Opera Software	*www.operasoftware.com*
Optimizing Apache's overall performance and bandwidth consumption	The Apache Group	*www.apache.org/docs/misc/ perf-tuning.html*
Optimizing the bandwidth-related performance of Windows NT (as featured in Chap. 12)	Microsoft	*www.microsoft.com/syspro/technet/ boes/bo/winntas/technote/nt301.htm*
Optimizing the performance of Microsoft Exchange Server (as featured in Chap. 13)	Microsoft	*www.microsoft.com/syspro/technet/ boes/bo/winntas/technote/perfbench.htm*
Parallel File Transfer Protocol (PFTP)	Theory Center of Cornell University	*www.tc.cornell.edu/UserDoc/HPSS/ Bookshelf/UserGuide/userguide-6.html*
Performance issues related to the 1.0 version of the HTTP protocol	World Wide Web Consortium	*www.w3.org/Protocols/HTTP/1.0/ HTTPPerformance.html*
Performance issues related to the HTTP protocol	World Wide Web Consortium	*www.w3.org/Protocols/HTTP/Performance*
Pingroute, a UNIX network monitoring tool	Stanford University	*www.slac.stanford.edu/comp/net/ mon/tool/README-pingroute*
Protocols available under Windows NT	Microsoft	*support.microsoft.com/support/kb/articles/ q128/2/33.asp*
SNMP-related network tools	Avatar Consultants, Inc.	*www.avatar.com/snmp.htm*

Information on	From	Can Be Found at
Some of the math that underlies signaling theory and practice (Be aware that much of the material at this site is displayed in Korean)	Korea Advanced Institute of Science and Technology	*mgt.kaist.ac.kr/home/Lab/Heuristics/ Chapter4.html*
Switched internetworks for workgroups	Cisco Systems	*www.cisco.com/warp/public/729/General/ swtch_wp.htm*
The C- and Java-based Web browser and editor Amaya (Not only documentation, but also source code and precompiled nbinaries are available from this URL.)	World Wide Web Consortium	*www.w3.org/Amaya*
Trouble-shooting frame-relay circuits (general); the Hewlett-Packard network monitoring tool Internet Advisor	Hewlett-Packard	*www.tmo.hp.com/tmo/pia/InternetAdvisor/ English/br-whousing.html*
Updating TCP/IP under Windows for Workgroups	Microsoft	*support.microsoft.com/support/kb/articles/ Q122/5/44.asp*
UTP cable characteristics	Students of the Worcester Polytechnic Institute	*bugs.wpi.edu: 8080/EE535/hwk96/ hwk2cd96/yinlu/temp.html*
Windows NT's Remote Access Service and its processing of traffic	Microsoft	*support.microsoft.com/support/kb/articles/ q97/5/59.asp*

INDEX

ABOUT THE AUTHOR

Michele Petrovsky, an author on and teacher of computing topics, earned her MS in Computer Information Science from the University of Pittsburgh. She wrote *Implementing CDF Channels, Microsoft Internet Information Server Sourcebook*, and *DHTML in Action*.